Young Children
Learning
Mathematics

Young Children Learning Mathematics

Douglas E. Cruikshank
Linfield College
McMinnville, Oregon

David L. Fitzgerald
Temple University
Philadelphia, Pennsylvania

Linda R. Jensen
Temple University
Philadelphia, Pennsylvania

Allyn and Bacon, Inc.

Boston / London / Sydney / Toronto

Library of Congress Cataloging in Publication Data

Cruikshank, Douglas E   1941-
   Young children learning mathematics.

   Bibliography: p.
   Includes index.
    1. Mathematics—Study and teaching (Preschool)
2. Mathematics—Study and teaching (Elementary)
I. Fitzgerald, David L., joint author. II. Jensen, Linda R., 1949-   joint author. III. Title.
QA135.5.C78   372.7   79-18154
ISBN 0-205-06752-2

Printed in the United States of America.

Photographers:

Douglas E. Cruikshank
William M. Sheffield

The term Cuisenaire rods appears frequently throughout the text. The name Cuisenaire® and the color sequence of the rods, squares, and cubes are trademarks of the Cuisenaire Company of America, Inc., and are used with its permission.

Contents

Chapter **8** **Preparing Children's Learning Environments** **274**

Preface

Young Children Learning Mathematics is designed to assist those people working with children, ages three to nine, in presenting early mathematical ideas. The text outlines the children's world of mathematics, how children develop mathematical understanding and mastery, and how an adult working with children can enhance the learning environment. Children begin their mathematical development long before they enter school. Mathematical learning begins as soon as infants first discover the relationships that constantly surround them. Mathematical learning builds throughout early play and the informal approaches of the preschool and kindergarten. The foundation of mathematical learning should culminate in success during the early elementary school years. This text brings to the reader the rationale and techniques for developing a sound base for mathematical learning, from preschool through the middle elementary school years.

For the past twenty years, mathematicians, mathematics educators, learning theorists and, primarily, teachers have developed mathematics projects and programs, affected textbook revisions, and introduced new teaching approaches. *Young Children Learning Mathematics* represents a distillation of the modern mathematics into a practical guide for prospective teachers and practitioners. It includes the application of psychological principles to the teaching and learning of mathematics. The wedding of mathematics and psychology is in its infancy in application. What mathematics children should learn and how they should learn mathematics sometimes conflict in classroom practice. We have organized these ideas and present them by sequence and example.

The beliefs on which this text are founded emanate from study, involvement, observation, research, and practical experiences. Each belief is listed and briefly discussed below. Making the transition from beliefs to application is of primary importance: this is the objective of *Young Children Learning Mathematics*. It is with the preschool and elementary teacher in mind that these beliefs are presented.

1. Children and adults are the most important ingredients of any learning experience. Learning mathematics takes place as parents, friends, peers, and teachers interact with youngsters. How well children learn mathematics depends on the quality of their interactions related to mathematical concepts. Children need encouragement and understanding. Above all, children need trust and support.

2. Ideas from Piaget, Skemp, Bruner, and Dienes, among others, have been instrumental in fashioning a practical approach to developing mathematical foundations in the early years. How children develop mathematical concepts is crucial in designing learning sequences for teaching mathematics. The mathematics that can and should be taught at any given time must be tempered with what is known about children and their developmental patterns.

3. The learning styles of children should be accommodated by the learning environment. When it is obvious that children need manipulatives to develop or reinforce a concept, then manipulatives should be used. As children gain understanding and require structure to develop or reinforce a skill, structure should be established. We have presented structured work with concrete objects, activities for developing concepts, and many ways to help children memorize as well as understand important mathematical information.

4. A sequential presentation of mathematical concepts and skills is only part of early mathematical instruction. In every classroom, there is dynamic interaction among students and adults. Management of a class, types of physical objects, acquisition of textual materials, use of materials—all need to be taken into account in the successful development of mathematics.

5. A mathematical concept is abstracted when it is identified as the common element among several apparently different embodiments of the concept. Once abstracted, the concept should be reintroduced regularly as it becomes the focus of a lesson or the foundation for further learning. As it is reintroduced, the concept is expanded to add further meaning and elaboration. The growth of mathematical understanding relies on this *spiral* presentation. *Young Children Learning Mathematics* directs the teacher to develop mathematics by employing such a spiral approach.

6. Common elements cross all elementary school subjects. When those elements are employed in presenting two or more subjects simultaneously, the integration of subject matter takes place. Integration of mathematics and language arts, science, social studies, physical education, and art is encouraged. Besides allowing for common aspects of various subjects to be identified, integration provides a learning environment that parallels learning patterns in life itself.

Acknowledgments

We hope our readers are the finest teachers ever to assist children in the learning endeavor. We need more skillful, knowledgeable, and compassionate individuals as teachers. If this textbook can, in any way, help prepare, direct, and encourage teachers in nurturing the mathematical growth of children, then the writers will be fulfilled.

Preparing this manuscript has been a new adventure for each of us. We have been constantly encouraged and supported by Steven Mathews from Allyn and Bacon, Inc. We extend our heartfelt appreciation for his assistance. A textbook about children is incomplete without contact with children. We want the children in three communities in particular—Avon School in Barrington, New Jersey; Northeast Philadelphia; and Lower Bucks County, Pennsylvania—to know how they have inspired us. Their patience with us as teachers and photographers is commendable. A special thank you goes to Miss Linda Deininger whose calm persistence resulted in a superbly typed manuscript copy. Finally, our families and friends deserve much credit for their understanding and support. We are deeply grateful.

D.E.C.
D.L.F.
L.R.J.

Young Children Learning Mathematics

Children and Their Mathematics

The Children's World

Young children are natural learners. Their potential and energy for learning is unmatched by any other age group. They search for meaning and explore their world. From their very beginnings, young children develop in a world quantified by previous generations of scholars. Space, number, shapes, puzzles, time, distance—all provide a rich milieu in which to grow. Providing experiences for learning numerical, spatial, and measuring notions relies on teachers and parents knowing about the children's world. Several aspects of the children's world, which are intended to establish a foundation for the early learning of mathematics, follow.

Children Want Number Experiences

Children want number experiences because number has been a part of their lives from the moment they, as infants, began to communicate. Communication and physical movement have always included intensity of sound; varying duration of activity; exploration of space; embrace and separation; sequence of occurrences; similarity and differences of humans, objects, places, and emotions. These relationships are recorded by youngsters to provide the basis for early quantification of their world.

Children expect to succeed in their early number work. The study of mathematics is the study of relationships, or of how things *are connected*. It is natural, then, for children to expect success, because they have been actively experiencing relationships for years before they enter school. They do not have the fear of failure sometimes systematically developed by schools. The current thrust of extending children's preschool experiences into the early school years has increased their chances of developing their abilities to meaningfully quantify their world.

Children Are Active in Their World

Physical and mental activities characterize the young during nearly all of their waking hours. Teachers and parents rarely have to instruct children on how to be active or how to play. School is one of the first places where quiet, orderly groups of children are found. An unnatural passiveness begins the long road to conformity. The thread of play, which is the labor of the young, should be woven through the early school years.

Physically active involvement should characterize the children's work in mathematics. Materials such as stacking blocks, cardboard boxes, pattern blocks, construction bricks, colored cubes, logical blocks, tinkertoys, string and beads, puzzles, Cuisenaire rods (see Appendix), sand, clay, water, and various containers should be basic in the classroom. Children use these materials to motivate counting, developing patterns, creating, observing, constructing, discussion, and comparison. From the manipulations and observations come the abstraction of quantitative ideas and the communication of these ideas in pictorial or symbolic form.

Photo 1–1

*Children Constantly Observe Relationships
in Their World*

The world makes sense to children and adults as connections are made among pieces that seem separate at first glance. Language is often closely tied with the expression of relationships. Relationships may be simple. For example, by hearing a sound followed closely by attention or fondling, the infant begins to learn the relationships between his name and himself. Somewhat later, he sees that other individuals have names. He is likely not to distinguish between a person's name and the name of the position that person holds in the family constellation. That is, Linda and Lori are names, whereas mommy and sister, although used as names, state family relationships. It is later in the child's development that these interfamily relationships are understood. Still later, grandparents, uncles, aunts, and cousins are linked. This recognition of relationships is only one of many that children make as a result of their experiences with their physical and human environments.

As children begin to quantify their world, they become aware of arithmetical, spatial, logical, and collective relationships through their active participation with their natural environment. Later, manipulative materials in the school classroom afford an effective, although synthetic, basis for mathematical learning. Some relationships are obvious. Others are not, and it becomes necessary for teachers to help provide experiences to link the more subtle relationships. Eventually mathematics makes sense, because the learners understand how most of what they have been learning is connected. Learning how things relate is sharply distinguished from learning by memorizing many disconnected facts. In the former case, mathematics is presented as possessing a structure; in the latter case, structure is generally ignored.

Relationships can be expressed in visual form. In moving from purely concrete work using physical objects to more abstract work, visual representa-

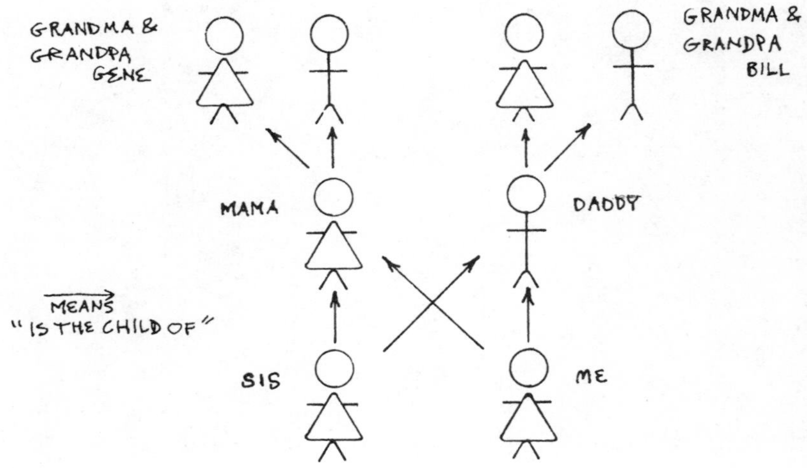

FIGURE 1–1

tions can effectively be employed. For example, Figure 1–1 is a visual representation of the relationships in one family drawn by a five-year-old. This representation links the concrete world to the abstract idea of family relations.

Children begin to experience mathematical relationships systematically when they begin school. Teachers should be aware of relationships and encouraged to develop experiences to further the learning of how things relate.

One of the more innovative contemporary mathematics programs, the Nuffield Project, involves children in an active learning approach based on the work of Swiss psychologist, Jean Piaget. Developers of England's Nuffield Mathematics Project have produced a chart that depicts the relationships among the many topics of their program. Figure 1–2 shows how various parts of the Nuffield Project are connected. It is a visual representation of more abstract ideas. (See p. 4.)

The Children's World Is Multifaceted

Infants cope with this complexity from the time of birth. Their early learning occurs in a milieu of interrelated occurrences and demands. The success rate for this period of early learning is remarkable. Thus, by the time children enter school, they are quite proficient in learning interrelated skills and concepts. Walking, talking, toilet training, rote counting, language development—each is learned without being isolated from other life experiences. Packaging bits and pieces of knowledge or separating skills from their applications defines artificial environments alien to the unified world of all humans. Integrated experiences in school are essential in providing more natural and balanced school experiences.

In teaching mathematics, the skillful teacher or parent of young children is afforded the opportunity to integrate that subject with other subjects and the world in general. The interrelatedness of mathematics with subjects such as art, language, literature, and science combine to help illustrate the close ties of all bodies of knowledge. As mathematics is embedded in other subjects, so also are other subjects embedded in mathematics. The educative process is one that brings many connected experiences to children in a lifelike environment. That children see the connections is crucial.

There are few thinking skills unique to mathematics. Rather, most thinking skills transcend specific knowledge or discipline. When taught in a setting of integrated learning, thinking skills provide the opportunity for children to learn how to learn. These skills of observing, describing, conjecturing, questioning, judging, valuing, communicating, to name but a few, are skills of life.

Children's Feelings Are as Important as Knowledge

Their feelings about themselves, others, life, education, or whatever topic, affect children's behavior. Knowledge divested of emotional impact does little to affect behavior. We can ill afford to treat children as if they did not matter. Certainly not all children will attack mathematics or anything else with the

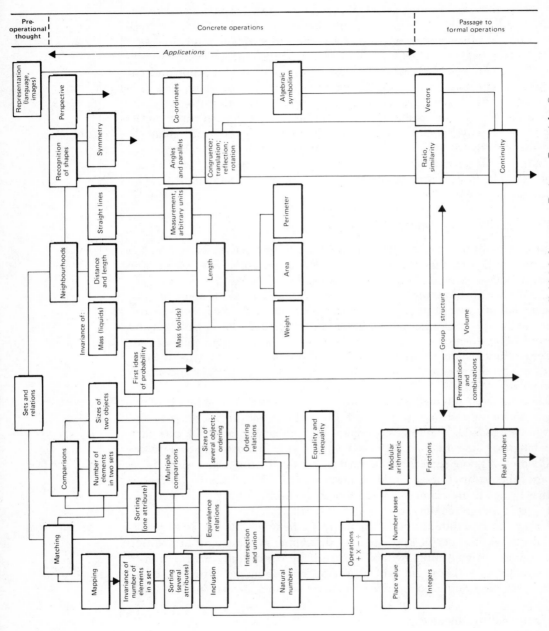

FIGURE 1-2 *Maths With Everything,* Nuffield Mathematics Project. Copyright © 1971, John Wiley & Sons, Inc. Reprinted by permission of John Wiley & Sons, Inc.

same energy and enthusiasm. Students should be exposed not only to the structures of mathematics but also to the historical and cultural aspects. Some students are interested in theoretical mathematics. Others are motivated by a historical approach. Still others will be excited by the relationships between mathematics and art, mathematics and music, or mathematics and language. Applications of mathematics will interest many children.

The feelings of students should play a leading role in elementary mathematics instruction. When teachers are not sensitive to the children's feelings, they tend to treat them like a collection of objects. Until children are treated as individual human beings rather than a collective product, they will not effectively learn any subject.

Children Thinking Mathematically

The overriding purpose for teaching mathematical skills and understanding is to develop a sense or feeling for mathematics. In the broadest meaning, learning mathematics serves as both a means and an end. Learning mathematics is a means to develop logical and quantitative thinking abilities. The key word is *thinking.* Thinking children are liberated from the dull routine that sometimes characterizes school. Learning mathematics is an end when children have developed basic computational skill and can apply mathematics to their world. That is, mathematics becomes functional in the lives of children. At least a part of children's environments can be explained because of simple mathematical principles.

Photo 1–2

At every level, the learning of mathematics should be a natural outgrowth from the children themselves. It should be interesting for the children. It should challenge their imaginations. It should beget creative solutions. It should be seen in their art, their dance, their music, their movement, and their conversation. Learning mathematics should be devoid of boredom, devoid of meaninglessness, and devoid of coercion.

Mathematics—Logical and Psychological Approaches

Approaches to teaching mathematics have generally followed the logical structure of mathematics. Thus, counting is followed by adding at the pictorial and symbolic levels. Then subtracting, multiplying, and dividing follow. Later, or concurrently, children learn the properties of these operations. To augment learning computation, sets, the study of fixed shapes, and simple measurement are included in courses of study. Understanding is developed to the extent children see meaning in what they are doing. Some children readily understand. Many others do not or cannot understand. The presentation of mathematics as an organized, logical structure does not assure understanding by young children.

To complement the logical structure of mathematics, the psychological aspects of learning mathematics must be weighed. Consideration of how children learn mathematics makes it possible for teachers and parents to develop activities that blend what is known about children and mathematics. Thus, developing an initial understanding of number involves classifying, relating, and ordering. Objects and groups of objects are used to illustrate and enhance the learning of number, the concept of an operation, and addition. Psychological considerations help children learn mathematics. The structure of mathematics helps children see how things are connected.

Sources of Information about How Children Learn Mathematics

The study of how children learn mathematics is not new, but until relatively recently, little had been written that was directly applicable for the classroom teacher. In the past few years, books and articles have appeared on the learning of mathematics. Among the most useful are the works of Piaget, Copeland, Skemp, Dienes, Bruner, and Gagné. Much of what appears in the following chapters rests on the foundation provided by these authors.

The reader is encouraged to independently explore how children learn mathematics. Only when teachers and parents begin to understand the aspects and stages of mathematical learning will children receive the kind of instruction most appropriate to their individual learning styles. In addition, such exploration will aid the reader to understand why this textbook places stress on manipulation of concrete objects, multiple embodiments of mathematical ideas, active participation of learners, alternative teaching strategies, use of mathematical relationships, and building mathematical ideas according to the developmental characteristics of children.

Mathematics education was particularly affected by Brownell, who set forth his "meaning theory" of arithmetic instruction in the 1935 yearbook of the National Council of Teachers of Mathematics. According to Brownell (1935, pp. 19, 31), " . . . this theory makes meaning, the fact that children shall see sense in what they learn, the central issue in arithmetic instruction." He went on to call for an "instructional reorganization" so that arithmetic would be " . . . less a challenge to the pupil's memory and more a challenge to his intelligence." In the ensuing years, general agreement has been reached among psychologists and educators that teaching which involves meaning or understanding tends to be richer and longer lasting. It has also been suggested that when learning is seen as a function of personal meaning, approaches to teaching must center on the children and their interpretations of what is being taught. Thus, teaching the meaning of mathematics provides an extra incentive for the teacher to know how children learn mathematics.

Children Form Mathematical Concepts

A concept is an idea or mental image. Words and symbols are used to describe or label concepts. For example, *potato* is a collection of sounds that brings to mind an image representing some generalized form of a garden vegetable. Exactly what image appears depends on the experiences, heritage, geographical location, and language of the listener. The symbol 5 represents a mental image of all groups containing ✳✳✳ ✳ or ◻◻◻◻ things. Again, the precise image that appears depends on the background and experiences of the listener.

Concepts are learned. Virtually all children from the time of birth have the capacity to learn concepts. Concept formation begins immediately. The language and symbols that name concepts lag behind concept formation but eventually emerge. As children grow and mature, language and symbols are introduced to name mental images already formed and, later, are used to teach new concepts. To learn a concept, children require a number of common experiences relating to the concept. Initially, a parent introduces potato to a child by spooning a white, strained substance into the child's mouth, perhaps exhorting the child to, "Eat your potatoes." As this procedure continues for several months, the child begins to associate the word *potato* with the mushy substance spooned into his mouth. Obviously, the concept of potato is very limited at this time. Soon, mashed potato from the parent's plate may be introduced to the child with the same plea: "Eat your potatoes." Over time, potatoes prepared in many ways are given to the child and, finally, after two or three years, the child is informed that the vegetable the parent is washing, peeling, and cutting is a potato and can be prepared in numerous ways. The concept of potato begins to emerge as an accurate, generalized mental image.

Two aspects of this example have clear implications for teaching young children. First, the concept of potato did not become *known* until it had been seen, felt, smelled, and tasted in many ways. That is, potato was introduced by having the child experience potato in numerous guises. When the child was able to discern the common property among the various ways in which potatoes were prepared, namely, each dish originated from a certain recognized vegetable, the concept of potato had been formed. Second, the word *potato* had to occur in concert or had to follow the experiencing of the vegetable. The sounds that make up the word, *potato*, would not have been at all helpful before the experience. The child would not have learned what a potato was, had he to rely only on hearing the word or a definition.

The two processes just described provide the foundation for the early learning of mathematics. Children who find the common property of several seemingly disconnected examples are abstracting. The abstraction that is made is a concept. Children learn their mathematics by abstracting concepts from concrete experiences. The language is developed during or after concept formation, never before. Objects and events that are a part of children's lives and are easily observed are less abstract than other objects and events not easily observed. Thus, dogs, automobiles, houses, toys, and mothers are less abstract than are color, height, time, and number.

Skemp (1971, p. 32) has stated two principles of learning mathematics that relate directly to the notion of concepts:

1. Concepts of a higher order than those which a person already has cannot be communicated to him by a definition, but only by arranging for him to encounter a suitable collection of examples.

2. Since in mathematics these examples are almost invariably other concepts, it must first be ensured that these are already formed in the mind of the learner.[1]

Children learn the meaning of number by experiencing number in many varied situations—through a suitable collection of examples. The same holds for learning addition, subtraction, multiplication, division, fractions, geometry, measuring, and so forth. The activities presented throughout the following pages are intended as typical examples for learning mathematical concepts. Also, building mathematical concepts requires constantly building foundations on which to base further mathematical learning. Attempting to develop mathematical concepts on a foundation of previously memorized notions, which were not understood, results in frustration for both children and teacher. Skemp (1971, p. 34) notes,

> . . . before we try to communicate a new concept, we have to find out what are its contributory concepts; and for each of these, we have to find out *its* contributory concepts; and so on, until we reach either primary concepts [derived from sensory and motor experiences] or experiences which we may assume as given.[2]

Dealing with Concepts Once Formed

The collection of concepts and experiences a person acquires makes up the knowledge that person possesses. As new experiences occur they are fitted

1. Richard R. Skemp: *The Psychology of Learning Mathematics*, (Pelican Books) p. 34. Copyright © Richard R. Skemp, 1971. Reprinted by permission of Penguin Books Ltd.
2. Ibid, p. 34.

into a person's existing mental structure. Depending on how familiar the experiences are and the learning style of the learner, the experiences may be easily received or rejected because of a person's mental structure or schema. The schema is a part of the mind used to build up the understanding of a topic. Thus, it is used to take in new ideas and fit them with what is already known to increase or alter what is already known. For example, as children are learning about potatoes and have experienced strained, mashed, and boiled potatoes, the schema of *potato* may be limited. When french-fried potatoes are introduced it may be difficult for children to immediately recognize the new food as potato. Although they are told what they see, smell, feel, and taste are potatoes, children may not be initially convinced that they are experiencing potatoes. The schema of potato must be changed to accept potato in its new form. Once this accommodation has taken place, that is, the schema has adjusted to now accept french-fried potatoes as potatoes, it may be said that the schema has assimilated the concept of french-fried potato. That is, french-fried potatoes are now understood. Understanding a concept means an appropriate schema has assimilated that concept.

The idea of a schema and how it functions provides a powerful tool for teaching mathematics. That a mental framework can be identified and developed means that mathematical relationships, patterns, and ideas can be understood rather than merely memorized. In the long run, children will have the ability to build up mathematical knowledge. When rules are memorized, children reach a point in their mathematical learning at which they are unable to remember the rules and are unable to continue learning. Understanding has long since vanished. As mathematical knowledge is introduced, its understanding is predicated on children having already developed appropriate early schemas. The implications are clear. Teachers should provide early mathematical experiences in a form that will assure that the mathematics is understood. Such a foundation provides for all later mathematical understanding.

Developing schemas provides for the future when memorizing rules cannot. What must be done so children can face the future, when what is to be learned is as yet unknown? Skemp (1971, p. 53) responds:

> The first part of the answer would seem to be to try to lay a well-structured foundation of basic mathematical ideas, on which the learner can build in whatever future direction becomes necessary: that is, to find for oneself, and help one's pupils to find, the basic patterns. Secondly, to teach them always to be looking for these for themselves; and thirdly, to teach them always to be ready to accommodate their schemas—to appreciate the value of these as working tools, but always to be willing to replace them by better ones.[1]

Children's Thinking

A goal of the teacher of young children is to provide experiences so children can progress from a concrete, intuitive level of learning dependent on the

1. Ibid., p. 53.

teacher toward a less concrete, reflective level of learning independent of the teacher. This is a life's work for both learner and teacher. Intuitive thinking is not a stage through which an individual passes on the way to reflective thinking. Even when an individual has reached the Piagetian stage of formal reasoning, intuitive thinking may be necessary to gain a preliminary understanding of a new, abstract concept. For a child yet unready for reflective thinking, intuitive thinking is the primary source of learning.

Learning by intuitive thinking means learning by experimenting with concrete materials, through experiencing ideas in various concrete ways, and by visualizing ideas without relying on analytic thought processes. Intuitive thinking allows concepts to make sense before full understanding takes place. For example, children who construct patterns using only Cuisenaire rods of two colors and stumble on the pattern below are not learning much about the commutative property of addition, but they are certainly gaining an intuitive grasp of what the commutative property means. Later, when the commutative property of addition is presented, these children should be able to understand the property. The bulk of children's early learning takes place at the intuitive level. Teachers have the responsibility to encourage such thinking.

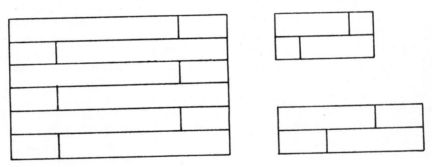

FIGURE 1–3

Reflective thinking comes later. Reflective thinking means being able to reason with ideas without needing concrete materials. The processes of reflective thinking include reflecting, inventing, imagining, playing with ideas, problem solving, theorizing, and generalizing. Reflective thinking allows individuals to know how something is accomplished rather than merely being able to perform a task. That is, many seven-year-old children can add two numbers, such as 18 + 17, but few can effectively explain how they accomplished the task.

Reflective thinking also allows individuals to alter or correct schemas. They may consider how an idea or process was perceived and compare it with a new interpretation of the same idea or process, leading to an alteration of the original schema. Piaget has noted that children develop the ability for some verbal thought with sufficient concrete representations by the age of seven or eight. From seven or eight to eleven, children have the ability to manipulate concrete ideas in their heads, but complete facility for reflective thinking comes later as the child approaches adolescence. Many adults never perfect their ability to think reflectively.

Skemp (1971, p. 67) has synthesized the role of the teacher to meet the goal mentioned earlier: to help children progress from a concrete, intuitive level of learning dependent on the teacher toward a less concrete, reflective level of learning independent of the teacher:

> He must fit the mathematical material to the state of development of the learners' mathematical schemas; he must also fit his manner of presentation to the modes of thinking (intuitive and concrete reasoning only, or intuitive, concrete reasoning, and also formal thinking) of which his pupils are capable; and finally he must be gradually increasing their analytic abilities to the stage at which they no longer depend on him to pre-digest the material for them.

Children Communicating Mathematical Concepts

Young children think quantitatively long before they engage in their first school activities. They have explored their personal space and they begin to think about the proximity of objects. That is, children become aware of objects near to them and those farther away. They notice fingers are close to a hand or arm, eyes are near the nose, and that grandparents live far away, even if they live down the street.

Order is another spatial relationship about which children think. They may have noticed the order of the cars on the toy train in their crib; or they may be aware of the sequence of significant events when they cry, such as a parent appears and they are held and comforted. Children classify objects as belonging together or not belonging together; for example, close family members versus neighbors and friends. They begin to judge objects as being few or many, big or small, tall or short, fast or slow. Obviously, they are not studying mathematics *per se*. The children are, however, thinking about their world quantitatively.

As children experience quantitative events and develop language to express these ideas, they are able to communicate with other people. They are developing the ability to classify objects and events more precisely. The language that emerges may not sound mathematical. Regardless, it does represent the foundation on which the more exact language of mathematics is built. When children discriminate by volume, they might use these "volume" words:

much	full
lots	empty
some	little
more	huge
all	less

1. Ibid., p. 67.

When discriminating by size, children might use the words:

big	thin
little	fat
tall	fatter than
short	long
biggest	wide
bigger than	widest
smaller than	

When indicating time, children may use:

before	last summer
now	tomorrow
after	spring
later	winter
when the bell rings	when it gets warmer
yesterday	

When discussing the location of objects, children may use:

here	inside
there	outside
up	above
down	below
on top of	around the corner
over	in the box
under	

Children's language develops in concert with their experiences. The experiences are crucial for the language to make sense. Likewise, in the initial stages of mathematics learning the quantitative experiences must be closely connected to the language that describes those experiences. A very serious mistake occurs when preschool and elementary school children are exposed to the language before or without any physical experiences to act as models, from which concepts can be abstracted. A common example is teaching addition before children know what an *addition* is; that it is an operation, that the operation transforms one number to another, that there is but one quantity resulting from each transformation, although there are many ways to express that quantity, and that addition is an extension of the idea of joining a group of objects with another group of objects using an analogous operation.

It is the teacher's responsibility to provide experiences and the concomitant language that allow for a natural wedding of experience and language. A workbook or textbook alone cannot perform this function. Children must be physically active. The language emerges and develops informally. Language development is enhanced as the teacher has frequent discussions with individuals, small groups, or the entire class about a particular activity or discovery. Experience stories may be written by the teacher or children to describe quantitative experiences. This is a natural extension of mathematical and language growth.

To sustain a conversation and communicate clearly, an individual must use language patterns that fit previously established conventions. It is not so important what sound a word has, as long as once that sound has been made, both the speaker and listener have some common mental picture or notion that makes sense. Youngsters learning how to talk commonly make sounds describing objects that for them are perfectly clear. It is the listener, usually a parent, who cannot determine what it is that is being mentioned. Most often, a pantomime is enacted that either produces the object or describes it well enough for the connection to be made. It is common for this language barrier to be found among adults as well. For example, *donkey* and *pulaski* are terms that make sense to a forester but could easily be misinterpreted or unknown to an accountant. The forester might verbally describe what was meant by the terms whereas the accountant might conjure up mental images that may or may not be distorted. The forester might show pictures or drawings of objects represented by donkey and pulaski, in which case the mental images would be much sharper. On the other hand, the accountant might accompany the forester to places where donkeys and pulaskis were being used, and the accountant could develop mental images based on all of his senses.

Photo 1–3

Mathematics is an area of knowledge in which language often causes distortions and misunderstandings. The confusion most often occurs when the unfamiliar language is presented before the experiences described by the language. Once introduced, the language may be verbally explained whereby the children may conjure up mental images that may or may not be distorted. They may also decide not to attempt to develop images and cease to listen. They may simply memorize the verbal description and repeat it verbatim as a response to a stimulus. The language may also be introduced by drawings, pictures, or diagrams. Such presentations are generally more helpful for children; however, the illustrations sometimes make little sense.

By way of contrast, another way the language may be introduced is after children have experienced the ideas that the language describes. Thus, children may discover that by using simple machines they can develop an understanding of the term *operation*. For example, consider a machine with an *input*, *operator*, and *output* that accepts as inputs various logic blocks, and whose rule for operating is to change the color of the incoming block. A red block placed

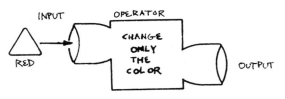

FIGURE 1-4

in the input may become either blue or yellow in the output, that is, after the operation has been carried out.

Another machine might be a "join" machine, in which a group of objects is placed in the input, and the operator might cause the child to join three objects to the input group. In such a case, the output group would be the input group joined with the group described by the operator.

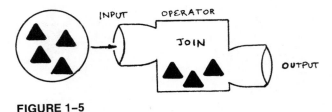

FIGURE 1-5

Many other machines could be devised to allow children to manipulate and discover the notion of operating or transforming. The term *operation*, when later applied to this transforming, has a good chance of being understood.

Language and symbols for language stand for abstractions that have been formulated in the minds of students. It is by language and symbols that we communicate what has been abstracted. In order to be meaningfully communicated, the abstraction must have taken place. Unfortunately, students are often introduced to language and symbols before they are ready to make the appropriate abstractions. To abstract a concept students need to do a variety of concrete operations. From these activities, which have a common structure, youngsters draw mental pictures or images. These images are what are common to the structures and are the abstractions of concepts to be learned. Once concepts are understood, they can be verbalized and symbolized meaningfully.

Adults' Beliefs about Children

Much of this book focuses on children as learners of mathematics, but it is important also to look just at the children themselves. The teacher's beliefs about children will affect the child's performance in many areas, including the learning of mathematics.

The beliefs adults have about children, how children should be treated, and how they should be taught vary dramatically. Some believe children must be left alone, to grow and develop with little interference from adults. Others believe children must be closely watched and directed. Surely the optimum treatment of children will include some combination of *laissez-faire* and strict direction. The overall degree of teaching success relies more on what teachers believe about children than on how they organize to teach them, because beliefs about children cause teachers to interact with children in ways that reflect the teachers' beliefs.

One example has been described in the research literature. When teachers believe children are low achievers, the children tend to receive marks indicating low achievement. When teachers believe comparable children are high achievers, the children tend to receive marks indicating high achievement. Teachers are often unaware that they treat children according to their personal beliefs about the children. They do so through both verbal and nonverbal interaction. As beliefs about children become more positive, a greater amount of the children's potential can be realized.

Many authors have discussed their beliefs about children. One such list is presented here to illustrate the kind of positive tenets that may ward off failure. Kelley (1965, pp. 7–15) listed and discussed twelve beliefs he held.

1. *Human beings are the most important things in the world.* . . . Every human being is an asset. To put it more narrowly, he produces more than he consumes. Far more important, he adds to *humanness.* . . . The person is more important than the textbook.

2. *Children are people.* . . . We often do things to children that we would not even consider doing to adults.

3. *Each person is unique.*

4. *When any human being is lost or diminished, everybody loses.* . . . Physical death is not the only form of death or of being diminished. When anybody is made to think less of himself, to feel less able, it is

partial death; and if it continues, the individual can become dead in the sense that he has become ineffective, immobilized, unable to enhance himself or others.

5. *Our children are all right when we get them.* By this I mean they are all right when they are born but not necessarily all right when they come to school. . . . More defectives are made after birth than before. . . . [W]hen a child behaves in a way detrimental to us and to himself, he has been made this way by the life he has been required to live in an adult-managed world over which he has no control. We (adults in general) have made him that way. His psychological self has been damaged, just as one's physical self may be damaged by disease or accident. If we could see the psychological self as we can the physical, our hearts would go out to these children. We are prone to pity the physical cripple and blame the psychological cripple.

6. *Every human being can change and change for the better as long as he lives.* . . . If it were true that nothing can be done for a child after the age of five, there would be little need for teachers. If teaching were confined to the dreary business of getting, people already ruined, to read, write, and cypher, followed by drilling the "facts," the teacher's life would be bleak. But when one realizes that the teacher has the opportunity to take damaged young ones and show them ways to growth and fulfillment, the task of teaching takes on new meaning and excitement. The possibility of change is the teacher's great reason to be.

7. *No one of any age does anything with determination and verve without being involved in it.* . . . We teachers are so accustomed to deciding what the tasks will be before we even see the learner that most of us lack skill in consultation. We are so full of our own compulsions and purposes that, even when we attempt to consult with the learner, we wait for what we had in mind to emerge before we register approval.

8. *How a person feels is more important than what he knows.* . . . This is true because one's feelings and attitudes control behavior while one's knowledge does not.

9. *Freedom is a requirement for humanness.* . . . A child, told exactly what to do, watched, tested to see whether he did it, and having sanctions applied when he does not or cannot do it, does not know freedom.

10. *All forms of exclusion and segregation represent the evil use of power and are evil.* . . . [I] include the segregation of bluebirds from crows; the exclusion of accessible people from school; the disenfranchisement of those whom adults frown upon from participation in the school council; the denial to some of the right to take part in student activities; the separation of children into X, Y, and Z groups to suit adult convenience.

11. *All forms of rejection are evil.*

12. *Our task is to build better people.* . . . It has been assumed that what lies outside the learner is what is important. That outside material,

often called *knowledge,* when injected, will of itself make better people. This has not worked, because too often the injection has called for too much coercion; and this has damaged the psychological self and reduced self-esteem. It is here that the individual learns hate instead of love.[1]

Kelley's message is simple: Teachers need to be keenly aware of the children with whom they work. A well-articulated set of beliefs provides the basis on which to develop a sound style of teaching.

Children's Learning Environments Are Enhanced by the Teacher

Teaching means directing, channeling, providing, suggesting, expecting, and encouraging children. Some people consider teaching an art; others consider teaching a science. It must surely be some of both. Skillful teaching is difficult. It is tiring. It is rewarding. Although much direction is given teachers in carefully designed teacher's guides to primary textbooks and workbooks, learning is improved immeasurably by extra effort. It is known how to be better teachers. There are several ways to become better teachers of mathematics.

Teachers would be better, if they enjoyed teaching mathematics. To do so requires appreciating mathematics. Children are not fooled by teachers who act as if mathematics is enjoyable, when it is not. Teachers must feel the same ecstasy as a youngster, who has discovered his first pattern or algorithm. They must feel the same confidence as the high school student after successfully completing his first geometric proof. They must feel the same satisfaction as the infant after his first steps. Teachers should allow themselves to become excited about mathematics.

Teaching would improve if teachers would allow themselves to learn more mathematics. Whether in formal classes or along with the children they teach, teachers can and should continue learning mathematical relationships, patterns, concepts, and skills. Students appreciate teachers who are obviously learning with them.

Mathematics teaching would improve if teachers would extend mathematics beyond the basal textbook. Teachers who are curious about what is happening in mathematics education will be immersed in the ideas of Dienes, Biggs, Davis, and many others who have gone beyond the textbook. The results of this exploration will provide intellectual fuel for thoughtful teachers.

Teaching would improve if teachers would seek out and communicate with school district colleagues who are excited and current in preschool and early elementary mathematics. Full advantage should be taken when colleague observations, in-service days, workshops, and conferences are offered in mathematics. Many teachers have been sparked by such experiences and have proceeded to become confident, competent, and dynamic teachers of a subject formerly thought to be beyond comprehension.

1. Paraphrased from Earl C. Kelley, ''What May We Now Believe?'' *Prevention of Failure,* pp. 7–15. Copyright 1965. American Association of Elementary—Kindergarten—Nursery Educators. Washington, D.C. Reprinted by permission.

Photo 1–4

Teaching would improve if closer attention were paid to how children learn mathematics and to the application of psychological principles that have emerged from the research of the past fifty years. The wedding of mathematics and psychology into the discipline of psycho-mathematics is exemplified by the work of Dienes and his associates throughout the world.

Goals of Mathematics Experiences for Children

The specific content of elementary school mathematics programs has traditionally been determined by authors of school textbooks. The inclusion and exclusion of particular topics has resulted from an evolutionary process. Educators and mathematicians meet constantly and exchange ideas. They are continually mapping out new directions. The expansion of mathematical topics in the late 1950s and early 1960s was, in part, responsible for the popularized "new math" movement. Of course, the content was not new to mathematics, but rather to teachers, parents, and students.

There is no clear consensus as to the appropriate content of elementary school mathematics. Tables of contents from textbook series are, however, amazingly similar. The content of instruction usually is an outgrowth of more general goals for elementary school mathematics. These goals stress developing a quantitative sense, thinking, communicating, appreciating, and understanding the structure of the number system as well as computational skill and problem solving. Most surely, these outcomes do not directly suggest a particular content to be studied. This is as it should be. It is far more important for children to experience thinking and communicating with mathematics as the medium than to cover an arbitrary content with thinking and communicating being chance outcomes. Thinking and communicating are skills that provide an individual with personal power, whereas knowledge of many isolated facts that have been memorized offers little personal power. *Power* here means what is useful to the individual when he interacts with others and needs the ability to think clearly, develop arguments, and generate knowledge. An isolated, memorized fact can be forgotten. Thinking allows it to be regenerated. It is, after all, the thoughtful, reflective individual that is needed; one who can be the master of mathematics rather than a prisoner of it.

Regardless of the content, teachers should accept the challenge to develop the more general goals of preschool and elementary school mathematics. Whether children are classifying objects, sorting attribute materials, employing ordering relationships, or developing fractional concepts, teachers should be encouraging thinking, communicating, generalizing, and other higher level skills.

Teaching Children Mathematics

Alternative Teaching-Learning Strategies.   Just as surely as there is unlimited human potential among individuals, there are also numerous learning environments. Within various learning environments, there are many alternative teaching-learning strategies. To elaborate, the learning environment consists of physical settings in which teaching and learning take place. Within the context of schooling, these physical settings are generally at or near the school building. Thus, the learning environment may be the physical organization of a classroom, the confines of a school yard, a nature trail, a creek, or an auditorium.

Within the learning environment, how a teacher behaves with respect to his or her learning objectives toward a particular group of children determines the teacher's beliefs, experiences, education, and so forth. In developing teaching-learning strategies, the teacher must be aware that many alternatives exist. Although never comprehensive, a list of some alternative teaching-learning strategies are carefully described in chapter 8. Each given strategy contains a single major focus, a single major type of student involvement and accompanying teacher behavior, and a single major type of student grouping. Figure 1-6 illustrates the three components of a teaching-learning strategy.

FIGURE 1–6

Developing any particular strategy is a unique function of an individual teacher and may only be successful under conditions experienced by that individual. Although it is recognized that there is no one *best* way to teach, teaching in only one way, that is, using but one strategy continuously, is not as likely to succeed over a period of time as is a variety of approaches. Teachers should be aware of various ways of teaching and learning. It will make teaching and learning more interesting and enjoyable. This is true of teaching mathematics as well as teaching reading, science, or social studies.

Some Issues Facing Teachers

After visiting classrooms to observe, participate with or teach children, prospective teachers or practitioners raise various concerns. Typical among comments is, "I saw the children playing a computation game and the boys were against the girls. The girls always seemed to win, and the teacher constantly praised the girls for being so fast. Competition is useful, isn't it? After all, life outside of school is competitive and children need to learn to compete early to survive." Other questions may focus on grouping children, on why and how children fail, and on how to manage or keep discipline in a classroom. There are no simple directions to avoid the pitfalls associated with teaching. A teacher's beliefs, upbringing, prejudices, values, self-concept, personality—these, and other characteristics affect an individual's teaching style. The approach a teacher takes to classroom organization and disruptions is an extension of who that teacher is. The quality of a teacher is a measure of the quality of the person who is a teacher. But what about competition, grouping, failure, and management? As examples of issues in teaching in general, as well as teaching mathematics specifically, each is briefly discussed.

Competition

A commercial advertisement promises consumers, "Being the best is not everything, it is the only thing." This challenge to competitors is part of the role expected in business. Seldom is the promise of being best realized, but competition is heightened. In the classroom of the young child stress on being the best, fastest, or brightest can prove destructive. For every "best" child in a group of twenty, nineteen are made to feel less able, weaker, or insignificant. Diminishing the worth of a single child diminishes the worth of the entire group. Employing competition to improve motivation or quality is ineffective. Only children who can win, compete; the others ignore the competition. The quality of a person's work emanates from a person valuing quality. A firm distinction must be maintained between being the best and doing one's best. This is not to say all competitive situations are harmful. Friendly competition sometimes results in increasing friendships and common appreciation among competing individuals or groups.

Cooperating and having a common cause foster personal growth and identification. Mathematics learning should be a cooperative effort. Coopera-

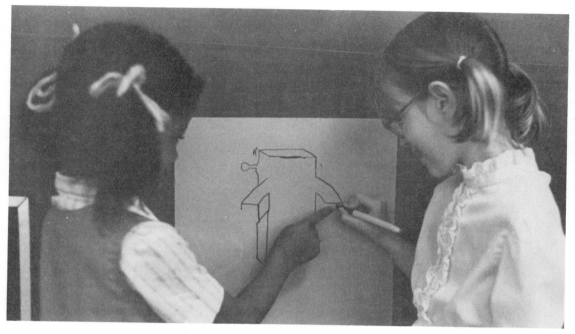

Photo 1–5

tion involves other people in constructive roles. Cooperation means that children can receive support from other people and can see the worth of working together as opposed to working against one another. But cooperation does not just happen; it is learned and requires a teacher who is a model of the cooperative spirit. More productive personal growth results through cooperation than through competition.

Grouping

The nature of young children does not lend itself to extensive group work. Typical four-year-olds are egocentric. When they play with one another, the interaction characteristically takes the form of parallel play. That is, children play in the same proximity of other children but rarely interact. They participate in independent activities. Sharing and cooperation is yet to be learned. There are group activities such as following the leader, listening to stories, and rest time. These activities are in the form of groups of individuals and play an important role in developing the notions of sharing and cooperation. If a group of four-year-olds sit around a table to work with colored cubes or Cuisenaire rods, their individual activities will be quite independent of one another.

Six-year-olds also have not as yet developed full capacity to work in groups. They do enjoy games involving several other children, but games that allow considerable individual freedom. When class projects involving all children are undertaken, the teacher is the catalyst and leader. Children will

Photo 1–6

sometimes assume group leadership in small groups, but rarely for the entire class. Opportunities arise to teach sharing and cooperation by developing small group projects, such as dramatic play or mathematics activities with a short-term, definite purpose.

Eight-year-olds have developed sufficiently to work in small groups with little or no adult leadership. They are able to grasp tasks and work them through. They are able to respond to the teacher's questions and guidance. They are learning to cooperate with one another without constantly grappling for attention. Care must be taken to teach skills of democratic living, and time must be provided so these skills may be practiced. This is the beginning of a lifetime learning process in group interaction. Eight-year-olds will be able to assume leadership in small groups but will rarely lead the entire class.

Grouping children by ability, need, or interest for instructional purposes is a common practice in early elementary grades. It is probably most common to group for reading, but it can also be an effective technique in mathematics teaching and learning. This practice is different from the type of grouping discussed above.

Grouping to facilitate instruction is an administrative technique to draw together children of similar ability or need or interest to expedite teaching, thus saving time and the effort to reteach material several different times. Care should be taken to consider children's abilities to work in group situations and the effect of a particular grouping pattern. Children constantly included in a low group soon perceive themselves as slow, dumb, or unworthy. Balanced grouping patterns should be employed so that through various mixtures of youngsters sharing and cooperation will take place in a variety of ways.

Failure

Young children seem to cope naturally with failure. The struggle to gain an initial understanding of life, including learning about the environment and roles within the family and society, involves trial and error. Failure in this context is used as a springboard to future success. It is the healthy use of mistakes. Failure can provide a foundation for growth.

Photo 1–7

On the other hand, failure may be devastating to the child. When teachers or parents teach that failure is bad and should be avoided at all costs, children develop sophisticated mechanisms to avoid failure. Holt (1964) discussed many reactions or strategies that children develop to avoid failing. For instance, some children excitedly wave their hands, when they really do not know the answer but know the teacher will call on those children who look apprehensive. Other times, children mumble an answer knowing the teacher will assume the right answer was given and will amplify on the mumbled response. At still other times, children will begin to give a response and then closely observe the teacher for clues, usually nonverbal, that the response they had begun was correct. Teachers may feel it is their duty to point out mistakes. Holt and others appropriately warn of the consequences of such behavior. It is far healthier to look for children's strengths and assets. Incorrect mathematical concepts cannot be ignored, but children can be guided to learn sound concepts without destroying their egos by condemning mistakes.

Children need to feel secure. The following are several suggestions for avoiding failure and developing children's abilities so they might be confident learners. Teachers and parents should:

1. Accept their own mistakes openly in front of children. Show children how failure is useful in learning and unimportant in determining one's overall worth.

2. Allow for failure and treat it as a natural part of the learning process. Make mistakes. Learn from mistakes. Avoid embarrassment as a result of failing.

3. Provide opportunities for children to succeed. Develop children's confidence through success to the extent that they will maintain confidence when they do not succeed.

4. Diminish the importance of doing everything "right." Avoid blaming children who are consistently unable to meet adult expectations. Expect a high quality performance within the capabilities of the youngsters.

5. Accept children as individuals worthy of respect. Believe that children deserve respect. Listen to children when they wish to share ideas and feelings. Respond in a warm manner.

6. Provide opportunities for children to make decisions that affect their lives within the school setting. Make sure the decisions are "real" and the children are willing and able to abide by the natural consequences of their decisions.

7. Discuss failure individually and as a class group. Come to agreements about how failure will be dealt with. Live up to those agreements.

Management

Management or discipline is of utmost concern to teachers and parents regardless of the age of the children. The factors that affect children's behaviors are many and varied. Some emerge from the very chemistry of the organism, most are learned from the home environment, others result from the school experience, and still others emerge through interaction with society at large. When groups of young children are brought together for instruction, the interaction produces individual and group behavior that must be constructively channeled. Even free play activities involving several individuals in parallel play must be managed so that children with overlapping interests avoid conflict. Cooperation and compromise must be introduced and nourished to the degree that young children understand those concepts. Even though ego-centeredness is characteristic, the rights and respect for others should be fostered. As Dinkmeyer and Dreikurs note (1963, p. 27), "Training of the child can be effective in many ways in a democratic setting. Simply stated, it must include a respect for order, avoidance of conflict, and encouragement." [1]

It is impossible in a short space to develop an ideal management or discipline plan for those people working with young children. Volumes have been written on the subject. Three such volumes are by Dreikurs, Ginott, and Dreikurs and Pepper. The reader is urged to select one of these or other books on classroom management and carefully incorporate in his or her teaching style management techniques with which he or she can be comfortable. It is hoped such techniques will be characterized by an ongoing respect for youngsters and their talents.

Photo 1–8

Encouragement

The processes of encouragement have been carefully studied. Self-evident as it may appear, encouragement is little recognized as a crucial factor in the

1. Dinkmeyer, Don, and Dreikurs, Rudolf. *Encouraging Children to Learn: The Encouragement Process.* Copyright 1963, Prentice-Hall, Inc., Englewood Cliffs, N.J. Reprinted by permission.

growth and development of healthy young minds. Teachers and parents tend to discourage youngsters even while professing the need for encouraging them. Discouragement emerges as one of the greatest single causes for failure. In addition to recognizing children's talents and weaknesses, adults need to support them through encouragement. Dinkmeyer and Dreikurs (1963, pp. 124–125) comment on the importance of encouragement in working with children:

> We, as educators, as parents, and as teachers, are in charge of the greatest treasure society possesses, the next generation. The urgent question which confronts us today is whether we will be able to guide them into becoming capable and responsible human beings or whether we will have to wait until youth itself claims its right to proper guidance and education. This question will be decided, in our opinion, by our ability to change from a punitive, retaliatory, and mistake-centered educational practice to one of encouragement for all those who have failed to find their way toward fulfillment.[1]

As a result of their extensive work on the encouragement process, Dinkmeyer and Dreikurs (1963, p. 50) have delineated nine methods of encouragement. The teacher or parent who encourages:

1. Places value on the child as he is.
2. Shows a faith in the child that enables the child to have faith in himself.
3. Has faith in the child's ability; wins the child's confidence while building his self-respect.
4. Recognizes a job "well done." Gives recognition for *effort.*
5. Utilizes the group to facilitate and enhance the development of the child.
6. Integrates the group so that the child can be sure of his place in it.
7. Assists in the development of skills sequentially and psychologically paced to permit success.
8. Recognizes and focuses on strengths and assets.
9. Utilizes the interests of the child to energize instruction.[2]

It is important to keep in mind the points raised in this chapter as the following chapters are explored. The child should remain the focal point as the mathematical ideas are developed.

1. Ibid., pp. 124–125.
2. Ibid., p. 50.

Extending Yourself

1. Become sensitive to the quantitative behavior of young children. Arrange to spend two or three periods observing children in a nursery, kindergarten, or primary grade classroom. Watch the movements of an individual or two. See how they explore space, solve problems, construct, sort, discover relationships, visualize patterns, and communicate.

2. Browse through a preschool, kindergarten, or first grade children's mathematics textbook or workbook. Notice how the author has sequenced the learning for the level of the children. Notice also the amount of space allotted for pictures as opposed to symbols. See how the author presents an idea and several pages later returns to reintroduce and extend the idea.

3. It has been said that most, if not all, mathematical experiences in the nursery school and kindergarten should be concrete. Defend or refute this statement by what you have read and experienced about children and their learning.

4. The process of abstracting a concept is crucial in learning mathematics. Develop a sequence of events that you believe would lead to abstracting the concept of "triangle." You might wish to parallel the description in this chapter of how the concept of potato is abstracted.

5. Describe in your own words the teaching implications for young children of Skemp's two principles of learning mathematics:

   (1) Concepts of a higher order than those which a person already has, cannot be communicated to him by a definition, but only by arranging for him to encounter a suitable collection of examples.
   (2) Since in mathematics these examples are almost invariably other concepts, it must first be ensured that these are already found in the world of the learner.

6. Consider Kelley's twelve beliefs about children. What do *you* believe about children? Do you agree with all of Kelley's beliefs? How would you alter them to fit your beliefs? Which would you delete? What additional beliefs would you list?

Bibliography

Association for Supervision and Curriculum Development. *Perceiving, Behaving, Becoming: A New Focus for Education,* 1962 Yearbook. Washington, D. C.: ASCD, 1962.

Brownell, William A. "Psychological Considerations in the Learning and the Teaching of Arithmetic." *The Teaching of Arithmetic,* The National Council of Teachers of Mathematics, The Tenth Yearbook. New York: Bureau of Publications, Teachers College, Columbia University, 1935, pp. 19, 31.

Bruner, Jerome S. *On Knowing.* New York: Atheneum, 1962.

_____ . *The Process of Education.* Cambridge, Massachusetts: Harvard University Press, 1977.

Copeland, Richard W. *How Children Learn Mathematics.* New York: Macmillan Publishing Co., Inc., 1974.

Dienes, Zoltan P. *Building Up Mathematics.* London: Hutchinson Educational, Ltd., 1960.

_____ . "An Example of the Passage from the Concrete to the Manipulation of Formal Systems." *Educational Studies in Mathematics.* Dordrect-Holland: R. Reidel Publishing Co., 1971, pp. 337–352.

Dienes, Zoltan P., and Golding, E. W. *Approaches to Modern Mathematics.* New York: Herder and Herder, 1971.

Dinkmeyer, Don and Dreikurs, Rudolf. *Encouraging Children to Learn: The Encouragement Process.* Englewood Cliffs, New Jersey: Prentice-Hall, Inc., 1963.

Dreikurs, Rudolf. *Psychology in the Classroom.* New York: Harper & Row, 1968.

Dreikurs, Rudolf, Grunwald, Bernice B., and Pepper, Floy C. *Maintaining Sanity in the Classroom: Illustrated Teaching Techniques.* New York: Harper & Row, 1971.

Gagné, Robert M. *The Conditions of Learning.* New York: Holt, Rinehart and Winston, Inc., 1965.

Ginott, Haim. *Teacher & Child.* New York: Macmillan and Co., 1972.

Holt, John. *How Children Fail.* New York: Pitman Publishing Corporation, 1964.

Inhelder, Barbel and Piaget, Jean. *The Growth of Logical Thinking from Childhood to Adolescence.* New York: Basic Books, Inc., 1958.

Jersild, Arthur T. *Child Psychology.* Englewood Cliffs, New Jersey: Prentice-Hall, Inc., 1964.

Kelley, Earl C. "What May We Now Believe." *Prevention of Failure,* American Association of Elementary-Kindergarten-Nursery Educators, Washington, D.C.: National Education Association, 1965.

Nuffield Mathematics Project. *I Do, and I Understand.* New York: John Wiley and Sons, Inc., 1967.

_____ . *Maths With Everything.* New York: John Wiley and Sons, Inc., 1971.

Payne, Joseph N., ed. *Mathematics Learning in Early Childhood,* Thirty-seventh Yearbook. Reston, Virginia: National Council of Teachers of Mathematics, 1975.

Piaget, Jean. *The Child's Concept of Number.* New York: W. W. Norton and Co., 1965.

_____ . *Science of Education and the Psychology of the Child.* New York: Orion Press, 1970.

_____ . *To Understand Is to Invent.* New York: The Viking Press, 1973.

Piaget, Jean and Inhelder, Barbel. *The Psychology of the Child.* New York: Basic Books, Inc., 1969.

Rosenthal, Robert and Jacobson, Lenore F. "Teacher Expectations for the Disadvantaged." *Scientific American,* Vol. 218, No. 4, April, 1968, pp. 19–23.

Skemp, Richard. *The Psychology of Learning Mathematics.* Baltimore, Maryland: Penguin Books, Inc., 1973.

Stroud, James B. *Psychology in Education.* New York: David McKay Co., Inc., 1956.

C H A P T E R

Children Learn Structure and Relationships: Toward a Concept of Number

The Early Basis for Understanding Mathematical Concepts

Relationships are rules or agreements by which we associate one object or abstraction with another. Structure is a system or pattern of relationships. As young children learn mathematics, they are analyzing, manipulating, describing, inferring, and inventing relationships and structures. For those people working with young children, it is crucial to understand how children develop mathematical understanding.

Children Experiencing Objects

Children begin to sense and observe their physical world at birth. Their usual first response is to assure doctor and parents of the quality of their lungs by screaming. They have for the first time begun to observe their world and react to it. They have begun to form crude relationships that will become increasingly more refined.

By the age of two or three weeks, children begin to demonstrate clear and conscious knowledge of their ability to use their senses selectively. They are able to fix their eyes upon an object and follow it if it moves. Gradually, they gain control of motor coordination so that they can turn their heads to look at an object. Soon they are able to respond to a sound by looking in the direction from which it came.

By two months, children have begun to develop systems of classification. They begin to classify experiences as pleasant or unpleasant. Taking a bath may be pleasant, whereas being changed may be unpleasant. Though they are clearly able to distinguish between experiences, their responses are probably reflexive rather than premeditated, conscious thought. For example, they may coo and smile during a pleasant experience, and yet have no knowledge that those reactions are communications usually associated with pleasant experiences.

As children have more experiences, they are able not only to observe and gather information with their senses, but also can relate current observations to past observations. They are recognizing. Recognizing is a simple but conscious form of classification. A relation or association is formed between the object observed and a past observation. They have put a new experience into the same category as a previous one.

Even children's earliest evidences of recognition may be considered crude mathematical thought. As children have additional experiences, they recognize a wider variety of stimuli and form more complex relationships among them. These relationships are gradually developed into an informal structure or classification pattern which becomes the basis for mathematical thought.

During this period of time, the children's environment should be rich in experiences that encourage them to use all of their senses. Such experiences allow children to develop more complex relational systems that accommodate a wider variety of stimuli.

During their first three years, children spend most of their waking hours observing their world and using their senses in a somewhat haphazard way. They observe a plastic object by tasting it. They gain information about a picture by touching it. Gradually, children make more sophisticated use of their senses by using the most appropriate sensory organ in a given situation.

Children Describing Objects

By the time children are three years old, they have begun to develop their most powerful tool for gathering information—language. They find they can often get more information about their surroundings by asking an adult than by sensing their surroundings directly.

By using language, it is possible to determine the level of development of children's observational powers and their ability to recognize objects they have already encountered. Children will begin to name objects, which they have encountered several times, if the people around them make a habit of mentioning the names to them. Parents or siblings commonly point to objects as they call the names. Children soon learn that pointing or naming an object is a sure way to draw the attention of those about them to the object. Slowly, from everyday life situations, children build up a vocabulary of the names of the objects they have observed.

Children should be encouraged to observe their surroundings and use whatever language capabilities they possess to describe their observations. It is important that children learn to observe objects in their environment and recognize the properties of the objects.

Some examples of activities which encourage children to observe and describe properties of objects in their environment follow:

1. Choose an object such as an item of furniture, a plant, or a magazine. Ask the children questions that elicit a discussion of the properties of the object. For example, what color is it? Is it rough or smooth? How is it different? Is it heavy or light? What is it used for?

2. Take the children to a special room of the school building, such as the kitchen or boiler room. Ask the children questions that elicit a description of the objects in the room and their possible function.

3. Take the children to a special area outdoors, such as the playground, a marshy area, or parking lot. Ask questions about specific characteristics of the area. For example, why is there no grass in the parking lot? Could a frog live here? Why are there no trees in the playground?

4. Buy or make a picture book of items that are in the children's immediate environment. Let the children look at the item pictured in the book and then point to the real item in the room. At this point, naming the item is not as important as recognizing it by its properties of size, shape, and color.

5. Have the children close their eyes and try to recognize everyday objects by the sound they make; for example, windows opening and closing, a light switch, a squeaky chair, writing on the chalkboard.

6. Choose a set of objects that can be identified by one of the senses acting alone; for example, the smell of peanut butter, the taste of a banana, or the feel of an ice cube. Deprive the children of their other senses by blindfolding, holding their noses, or covering their ears; then let them attempt to identify the objects by using only one of their senses.

7. Let the children fit geometric objects such as triangles and squares into frames of the same size and shape.

8. Make an outline board using outlines of everyday objects such as combs, clothes pins, and blocks. Have the children lay the objects in the correct position on the board. (See Photo 2–1.)

9. Make a find-a-shape activity in which the geometric shapes are hidden in a picture for the children to find.

10. Use feel-a-shape activities in which the children are asked to recognize a geometric shape by touch alone.

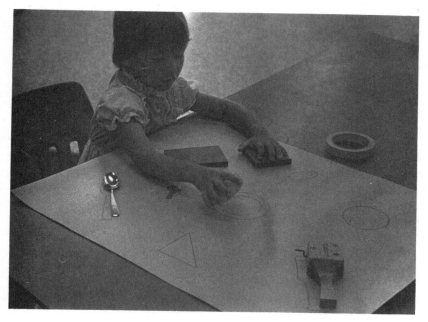

Photo 2–1

11.　Lay out a pattern of colored cubes and ask the children to reproduce the pattern with their own colored cubes.

Photo 2–2

It is more important at this stage that children observe and manipulate materials than it is for them to know the names of the material. Naming is not the same thing as knowing. Since naming is very easily evaluated, teachers often over-emphasize children's abilities to name objects accurately. Children may be able accurately to name objects with which they have not had sufficient experience. In addition, if an object is unfamiliar, naming the object may be impossible. However, it can still be observed and classified with other similar objects according to its properties.

Children who have an opportunity to experience numerous activities such as these soon realize they live in a world of relationships and patterns. The wallpaper in their room is an intricate pattern of repeated designs. Their toys have color patterns. The bricks on the front of the school building form an interlocking pattern. Moreover, they notice similarities and differences among objects. Some of their toys are large, and others are small. Some automobiles are alike, and others are different. Some dogs are mean, and others are friendly.

As experiences accrue, children begin to go a step beyond mere observation and attempt to draw inferences. For example, children will observe the wind blowing through the trees and conclude that the movement of the leaves causes the wind. This cause-and-effect reasoning should be encouraged at every opportunity, because it is an indicator of the onset of logical, deductive thought. It matters little that the reasoning and conclusions are often faulty. Inferential skill will become more precise with experience and practice.

When children first encounter a new object, they immediately manipulate and experiment with it. They begin to form inferences about the material and may draw conclusions about it that are amusing to adults. However, it must be kept in mind that children infer with immature and imperfect logical structures and limited information about their environment. Nothing should be done to discourage this natural tendency to hypothesize. It is better to let children form an incorrect conclusion than to correct their conclusions and discourage the tendency to hypothesize.

Children should be encouraged to describe objects by both sensing and inferring, but they should learn that there is a difference. In mathematics, a completely objective observation (equivalent to sensing) is usually in the form of a definition, axiom, or rule. What is inferred, such as conjectures or theorems, can be proved or disproved by additional information or a new organization of information that is sensed or supplied. Thus, the children should be encouraged to sense and infer, but they should know the difference.

Observation results from direct use of the sensory organs, whereas an inference is a guess about something that cannot be directly sensed. Children should be encouraged to infer, to question their inferences, and to try to develop a more certain method of checking their conclusions. Children's mathematical activities should allow many opportunities for guessing, questioning, blundering, and groping for answers. It is important for children to learn to recognize when a conclusion is certain and when it is still open to question.

Some activities which encourage inference and develop questioning skills are presented below.

1. Make a small *mystery box* into which several assorted, unrelated objects (marbles, buttons, flashlight battery, small cloth, and so on) have been sealed. The object of the activity is for the materials in the box to make noise but not be seen. You may wish to include one object that has an odor. Have the children observe and infer about the box.

Some possible observations are:

"The things inside make a noise."
"The things inside are heavy."
"I hear a metal-like sound."

Some possible inferences are:

"There is a spoon in the box."
"There is something made of metal in the box."
"Something in the box is round."

2. Ask the children if everything will fall. Let them try all kinds of objects ranging from bricks to sheets of paper to an inflated balloon. The children will infer that everything will fall unless something is holding it up. Discuss with the children why a bird does not fall.

3. Put attribute blocks (see Appendix) in a container so that the children cannot see them. Take them out one at a time, allowing the children to observe them carefully. Ask the children to guess which blocks may still be in the container. The children should not have had previous experience with the set of blocks used in this activity.

4. Start a pattern with Pattern Blocks (see Appendix). Ask the children to finish the pattern that has been started. There may be many correct answers, but children should be able to explain why they finished the pattern the way they did.

5. Have the children find things in the classroom about which they can both observe and infer.

Examples of observations:

"The air coming from this vent is warm."
"The window glass is cold."
"The table top is smooth."

Examples of inferences:

"The air from the vent is coming from the furnace."
"The air outdoors is cold."
"The table top has been sanded."

6.   On a windy day, go outside and let children observe and infer.

Examples of observations:

"We can feel the wind."
"It makes a noise in the trees."
"The wind is clear like glass."

Examples of inferences:

"The wind is really air because air is clear."
"It is air that is moving."

When children are observing and inferring, they should be encouraged to use descriptive words that are both comparative and noncomparative. Examples of comparative words are: large-small, rough-smooth, tall-short, thick-thin. Examples of noncomparative words are: plastic, glass, red. The teacher should use descriptive words frequently in their correct context, so children have an opportunity to refine their concept of the subtle meanings of the terms. Children often confuse the meanings of descriptive pairs such as more-less and large-small. It is common for children to think that a set has more objects in it because the set contains large objects. Children will also often claim that an object is taller, because it is higher.

Another frequent error made by children is using words that elicit emotions in the place of descriptive words. For example, children may say that an object feels "good" rather than observing that it feels "warm and soft." The teacher should pursue such statements and encourage students to name in specific terms the sensations that caused the "good" feeling.

Children Classifying Objects

Even before children learn accurately to describe objects in their environment, they begin to classify them. As soon as children become familiar enough with objects in their environment to recognize them, they categorize them into groups of other objects they recognize. The fact that children recognize and categorize an object does not imply that they should be expected to name it. Children may indicate that they have perceived a property of an object by simply discriminating between objects similar to the given object and those that are different. For example, children may sort square and circular objects into different categories without being able to name either shape.

Classifying is the process of grouping or sorting objects into classes or categories according to some systematic scheme or principle. Classifying is performed by using the specific, observable properties possessed by the object to be classified. That an object belongs to a class means it has some property in common with the other objects in the class. Classifying is a means by which individuals can cope with large numbers of objects by grouping them until they can see simpler relations in the complex assemblage.

Photo 2–3

The groups or assemblages into which objects are sorted are called sets. A set is a collection of objects, sufficiently defined, so that given any object it is possible to determine without question whether that object is in the collection.

Classification systems also serve to supply an accurate description of an object that is not available. By observing the position in a given classification system that an object would occupy, it is possible to describe the properties that the object would have. This is the fundamental process used in mathematical problem solving. A mathematical problem is a structural system that has a missing part or parts. By analyzing the system until the structure is understood, it is possible to describe accurately the properties possessed by the missing part. Inserting the missing part or answer into the structure completes the structure and solves the problem. Children should be given many activities in which they supply the missing part of a concrete structure. In their later mathematical experience, they perform the same process with abstract structures called equations.

Photo 2–4

Children's early experiences in classification should involve classification by only one property. The materials used should have only one obvious difference; for example, a set of beads that are alike except for color, or a set of sticks that are alike except for length. Though the objects should have only one difference to begin with, children should sort the widest possible variety of objects. Typical differences that might be used for children to sort are: shape, size, height, length, thickness, color, or material (that is, the substance of which an object is made, such as plastic or fabric).

Some examples of classification problems for children three to five years old involving only one difference follow.

1.   Give children a set of buttons that are alike except that some are red and some are blue. Ask the children to sort the buttons into two piles.

2.   Give children a set of household junk that consists of objects that belong either in the kitchen or the bathroom. Ask them to sort it into two piles.

3.   Give children a set of junk that consists of objects that are either metal or plastic. Ask them to sort it into two piles.

Photo 2-5

4.   Give children a set of toys that either has or does not have wheels. Ask them to sort it into two piles.

Children may think of a different way to sort the materials than adults do. Encourage this. Ask children to tell you how they are sorting the material, if they can.

5.   Give children a set of pictures of objects and have them sort the pictures into "know the name of" and "don't know the name of" piles.

6.   Give children a set of household junk and a small pan of water. Ask them to sort the junk into "float" and "does not float" piles.

As children become more proficient and are able to do more difficult classification activities, the number of attributes in the activities should be increased. There are several ways to increase the difficulty of activities. The difficulty of the task may be increased by: increasing the number of objects to be classified; increasing the number of categories into which the objects are to be sorted; increasing the level of abstraction of the categories, that is, happy versus unhappy rather than red versus blue.

An additional level of complexity that may be useful in children's later mathematical experiences is the problem of overlapping classes. For example, if children are sorting household items that include a plastic fork and the categories used are "things we eat with" and "things made of plastic," then the plastic fork will fit into both categories (see Figure 2–1). Children may need help denoting the answer in a clear fashion. The use of a Venn diagram of overlapping loops of yarn is probably the clearest way to solve the problem. (More discussion is in chapter 4.) Proficiency in denoting the concept of intersecting sets will become useful in children's later mathematical experience, when they study the concept of operations between sets.

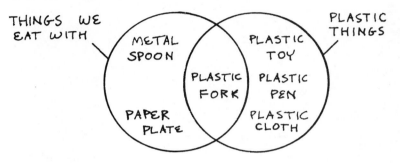

FIGURE 2–1

The following activities are more difficult classification activities, because of one or more of the reasons listed above.

1. Give children a large pile of household junk and have them classify it according to composition. For example: glass, metal, rubber, plastic, wood, leather, paper, fabric.

2. Give children pictures of animals and have them sort them according to habitat; for example, trees, water.

3. Give children pictures of people who are doing something and have the children sort them according to what they are doing. For example, the pictures might include a policeman, baker, librarian, nurse, mechanic, pilot, or professional tennis player.

4. Have children stand at the window where they can see automobiles and have them orally classify them by color, size, speed, and so on.

5. Give children a set of small squares of fabric and have them sort the fabric by checks, plaids, polka dots, or stripes.

6. Have children classify a set of toys according to whether they roll, make noise, are made of wood, and so on.

7. Have children classify objects in the room according to their function. For example, the objects might be to sit in, eat, or write with.

Photo 2–6

After children have acquired experience and expertise with classification activities like those above, they should be encouraged to invent classification systems of their own. They should be given sets of objects and asked to group them. The teacher should not specify what groups are to be used. It is important for children to understand there is no single correct answer for this activity. They should be free to invent whatever groups they want without fear of being judged on the basis of their choice.

The teacher can control the difficulty of the classification tasks in the same manner as before. At first, children should be presented with tasks that have few pieces to classify and a few concrete categories into which the pieces may be grouped. As the children's proficiency increases, the teacher may present them with a randomly assorted pile of junk and expect them to invent an appropriate classification system for it.

Examples of activities that allow students to invent their own classification systems are as follows.

1. Give the children a box of assorted buttons and ask them to classify the buttons. The assortment of buttons should be diverse enough to allow for a number of different classification systems including some that overlap.

2. Cut out pictures of people from magazines and mount them on posterboard. Ask the children to sort them in any way they like; for example,

mood—happy, sad; race—Black, White, Oriental; activity—sitting, standing, running.

3.   Make a collection of small objects such as pebbles, beans, beads, and coins. Ask the children to invent a classification system for the objects.

The discussion thus far has centered around the classification of objects that are found in their natural state in the children's environment. It is important that children have numerous experiences of this type. The children will tend to identify more directly with the activities, if they are done with materials with which they are familiar.

However, the natural environment does not always lend itself to the learning of the orderly structures of mathematics. Therefore, certain contrived materials that more accurately illustrate the mathematical structures are introduced into the children's environment. The materials have specific organized properties that bias the children's discoveries and enhance the probability that they will learn the desired mathematical structures. Such materials are usually called attribute or property pieces. Several examples of these materials are: People Pieces, attribute blocks, colored cubes, Parquetry pieces, and Cuisenaire rods. A description of these materials is given in the Appendix.

Children Abstracting Properties

The term *abstract* is often used as an adjective to connote a concept that is intangible or that is apart from the physical or material world. However, when the word is used as a transitive verb, it means "to take away" or "to summarize." Abstracting a property from a set of objects is the ability to look at a set of objects having varied properties and to identify the specific property that the objects have in common.

Abstracting may be described as the reverse of classification. In classification, a scheme exists into which an individual object may be fitted. Abstracting is the ability to look at the set of objects thus classified and identify the scheme by which the classification was accomplished. For example, children may look at a set of attribute blocks that vary in color, size, and thickness, and notice that the objects are all circles. This is a simple form of abstracting.

One error most commonly made when dealing with young children is to assume that the children fully understand any abstraction that they can name. Naming, however, does not imply a level of understanding. For children to use the word *chair* does not imply that they have abstracted the idea of chair. In fact, they may think that the only object called chair is the one in the living room that their dad sits in. Children may recognize the word, but may think that the word denotes only one object in the universe rather than a whole class of objects.

When trying to teach children a mathematical concept, it is absolutely essential that the children be exposed to large numbers of examples of the concept. In the case of the chair, the children must be shown wooden chairs, upholstered chairs, swinging chairs, folding chairs, reclining chairs, rocking chairs, and all other types of chairs that one can imagine, before it may be assumed that they have a complete understanding of the concept of chair. A complete understanding implies that children will be able to walk into a room in which they have never been before, see an object in the room that they have never seen before, and correctly identify the object as a chair.

This process of teaching an abstraction to children is known as the *multiembodiment principle.* The object of the principle is to embody a given concept in so many different forms, that children are able to abstract only the essence of the concept in such a way that their perception of the concept is independent of any specific concrete example of the concept. Their concept is then an abstraction in the sense that it is thought of as apart from material objects.

Abstracting, then, is the process by which we become aware of similarities or relationships between our experiences. Children should be provided with many activities that give them the opportunity to discover a relationship or similarity that exists among the objects in a set. Some examples of activities of this type are as follows:

1. Select a small group of children from the class who are the same in some way; for example, wearing glasses, same sex, same color of hair. Ask the class, including the ones selected, to try and guess the relationship between the members of the group. The common attribute of the children should be an obvious and observable characteristic. The activity can be varied in difficulty by choosing an attribute such as "wears a dress." The group would be all girls, but some girls would not be in the group.

2. Assemble a small pile of assorted material. Set one piece aside and ask the children to find another like it. The children are forced to form some class into which the piece will fit and may find several other pieces in the pile that are related to the given piece according to the relationship they have invented.

3. Show children a set of pictures of everyday objects and ask them how they are all alike. For example, a picture of a tree, a dog, and a butterfly would all be in the class of "living things."

4. Give the children a set of attribute blocks that are all the same size but vary in shape, color, and thickness. Ask the children how they are all alike.

5. Place a set of attribute blocks in front of a group of children and set one block aside. Ask the children to find a block that is like the one you have set aside. They may choose one that is the same shape, color or (less likely) thickness. All answers are correct. You may also wish to ask the children to choose a block unlike the chosen one.

6. Make two sets of People Pieces, such as male and female. Do not use all of the pieces. Give the children the remaining pieces and ask them to place these pieces in the sets. Let the children tell you why they placed the pieces as they did.

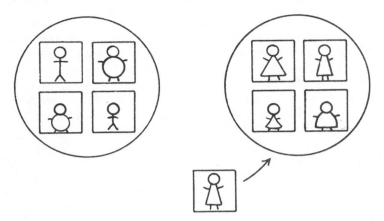

FIGURE 2–2

When choosing activities, be certain to choose a variety of materials with which the children may work. If the multi-embodiment principle is not used, then children may be able to readily identify a triangle when it is one of the attribute blocks but may completely fail to recognize it when it is drawn on the playground as part of a hopscotch game.

The Mathematical Basis for Exploring Relationships among Objects

Relationships are rules or agreements by which we associate one object or abstraction with another. In their early experiences, children tend to look at objects in isolation. They see one object at a time and even center their attention on one aspect of the object. As children become more experienced, they observe and interact effectively with two or more properties of an object simultaneously. Moreover, they become increasingly aware of how an object fits into the total environment in which it is found.

Mathematics is a system or pattern of relationships among objects or abstractions. It is important for children's early experiences to include numerous examples of relationships among objects. As children become more experienced and develop a wider repertoire of abstractions, they begin to look at relationships among the abstractions. Some of the relationships the children will explore are critical to their later understanding of mathematical topics. The most important of these relationships are order and equivalence.

Three Properties of Relations

When speaking of the properties of an object, one can give a complete description of the object by naming each of its properties. The more properties named, the more complete the description. The more complete the description, the less likely the confusion of the object with other objects in the environment. For example, if children are presented a set of attribute blocks and asked to select the red block, they are unable to do so because there are many red blocks in the set, and the definition has not been precise enough to allow them to choose the one the teacher had in mind. However, if the children are asked for the red triangle, they now have many fewer blocks from which to choose. But the description is still inadequate. To be complete, the children must be told some value of every variable (property) possessed by the blocks. The teacher must ask the children for the large, thick, red triangle. Now there is no ambiguity and no confusion. A precise definition of the block in question has been given by naming some value of each variable possessed by the block.

Relationships among objects possess properties, just as the objects themselves possess properties. Children should experience three specific properties of relationships: the reflexive property, the symmetric property, and the transitive property. These are properties of the relationships among objects, not properties of the objects themselves. As with objects, an accurate description of a relationship is given by telling whether or not the relationship possesses these three properties.

When speaking of mathematical relationships and their properties, two concepts must be clearly defined. One, the rule that defines the relationship must be clearly stated. Two, the set of objects or abstractions for which the rule is defined must be clearly stated. For example, consider the relation "is longer than" on the set of Cuisenaire rods. The rule states that rod A is related to rod B if rod A is longer than rod B. Thus, the orange rod is related to the yellow rod, the brown rod is related to the black rod, and so on. The relation here is being applied only on the set of Cuisenaire rods (see Figure 2–3). Therefore, when investigating the properties of this relation, only the set of Cuisenaire rods will be considered.

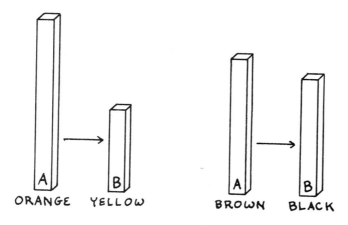

ORANGE   YELLOW   BROWN   BLACK

FIGURE 2–3

For this relation to be completely understood and classified with other relations, its properties must be investigated. *The reflexive property* states: *each member of the set would be related to itself by the rule of the relation.* Is each Cuisenaire rod "longer than" itself? Since the answer is clearly no, it is said that the "is longer than" relation on the set of Cuisenaire rods is not reflexive. An example of a relation on the set of Cuisenaire rods that is reflexive is the relation "is the same length as" (see Figure 2–4).

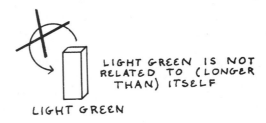

LIGHT GREEN IS NOT RELATED TO (LONGER THAN) ITSELF

LIGHT GREEN

FIGURE 2–4.   Reflexive property

The symmetric property states: *if rod A is related to rod B then rod B will also be related to rod A.* For example, using the "is longer than" relation, the brown rod is related to the yellow rod because the brown rod is longer than the yellow rod (see Figure 2–5). Is the yellow rod related to the brown rod?

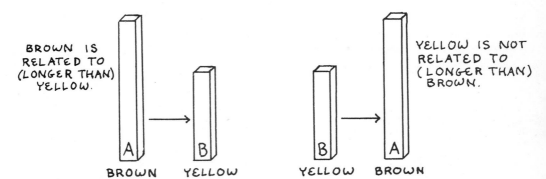

BROWN IS RELATED TO (LONGER THAN) YELLOW.

YELLOW IS NOT RELATED TO (LONGER THAN) BROWN.

BROWN   YELLOW   YELLOW   BROWN

FIGURE 2–5.   Symmetric property

Since the answer is no, it is said that the "is longer than" relation is not symmetric. An example of a symmetric relation on the set of Cuisenaire rods is the relation "is the same color as."

The transitive property states: *if rod A is related to rod B and rod B is related to rod C, then rod A will be related to rod C.* For example, using the "longer than" relation, the blue rod is related to the purple rod, and the purple rod is related to the red rod (see Figure 2–6). Is the blue rod related to the red rod? Since the answer is yes, it is said that the "is longer than" relation is transitive.

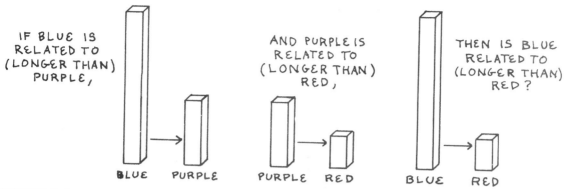

FIGURE 2–6.   Transitive property

Thus, an analysis of the "is longer than" relation on the set of Cuisenaire rods reveals that it is *not* reflexive and *not* symmetric, but that it *is* transitive. It would seem that this relation is not a very useful one, since it has only one of the three properties that a relation may have. However, an object is often useful, because it does *not* possess a specific property. For example, a dog may be useful as a pet for a small child because it is *not* mean. On the other hand, the dog may be useful for protection, because it *is* mean. Thus, it is as important to find that a relation is *not* symmetric and *not* reflexive, as it is to find that it *is* transitive.

Relations in mathematics that are transitive but not reflexive or symmetric are important enough to have their own special classification. They are called *order* relations. *An order relation on a set puts the members of the set into a sequence so each member of the set is assigned a specific and unique position in the sequence.* Thus, an order relation can organize a set in a very useful way.

The set of Cuisenaire rods is neatly sequenced by the relation "is longer than." The orange rod is first in the sequence since it is longer than all of the other rods. The blue rod is second since it is longer than every rod except the orange one. The sequence continues to the white rod which is not longer than any other rod and is, therefore, the end of the sequence.

In summary, to talk about relations and their properties, two mathematical concepts must be clearly defined. First, the rule for the relationship must be clearly stated. Second, the set upon which the relationship is defined must be clearly stated.

There are three properties of relationships:

1. Reflexive Property. If A is a member of the set, then A is related to itself.
2. Symmetric Property. If A and B are members of the set and A is related to B, then B is related to A.
3. Transitive Property. If A, B, and C are members of the set and A is related to B and B is related to C, then A is related to C.

Although young children may never know the names of the properties of relations, they should be exposed to examples of relations having a variety of combinations of the properties. The children's first experience should consist of very informal rules of comparison on only two objects at a time.

Some examples of activities of this type are as follows:

1.   Show the children two lengths of string and ask: "Which is longer?" "Which is shorter?"

2.   Let the children handle two balls of clay that are the same size but are very different in weight. (Put a lead sinker or a ping-pong ball inside one of the balls of clay.) Ask the children comparison questions using several forms of comparison statements; for example: Which is heavier? Which weighs less?

3.   Show the children two relatively unrelated objects such as a leather key holder and a can of pick-up-sticks. Ask the children comparison questions using several forms of comparison statements.

4.   Put two relatively unfamiliar objects such as a 35 mm film canister and an empty cassette case into an opaque bag. Allow the children to reach into the bag and touch the objects and encourage them to make comparison statements about the objects.

5.   Make a cassette recording of two unusual sounds, such as the operation of a garbage truck compressor and the sound of a harmonica. Let the children hear the sounds several times and encourage them to make comparison statements about them.

6.   Place two objects having pungent odors into opaque jars. Allow the children to smell the object in each jar and make comparison statements about them. The objects may be wrapped in a paper towel and placed into a clear jar. It is important that the children be able to smell the objects, but not to see them.

In each of the activities above, only two objects are given to the children at a time. This limits the complexity of the activity and encourages the children to make comparisons. It should not be inferred that only two objects should be used for each type of activity given. The children should have as many varied experiences as possible with as many different materials as time allows.

Ordering Relations

Ordering relations are relations that are not reflexive and not symmetric, but are transitive. Relations that order sets of objects are learned in an informal way by children at an early age. Ordering words such as "bigger," "less," and "taller" are used frequently by children in everyday conversation. A typical form of relational words used by children are words that denote present, past, and future events. Children will invent words such as "comed," "goed," and "hitted," which indicate that they understand and have generalized the use of the suffix "-ed" as meaning something that happened in the past. The children have probably never heard the words used by anyone else. They have abstracted the "-ed" as representing the past and have used the abstraction to form words that communicate in a way that is completely new. The children are not mimicking, but are inventing.

The process of ordering is a more complicated form of comparing. Comparing is forming a relationship between two objects. Ordering is a relationship between an object and a set of objects. An order relation applied to several objects allows one to take each object of the set and place it in a unique place in a sequence determined by the relation.

Ordering may be considered a type of sophisticated classification in which each object is in a class by itself. Moreover, the class into which the object falls is determined by the properties of the other classes before and after it. In the case of number, each number finds its place in the sequence because it contains the class just preceding it and is contained by the class just following it. This fundamental property of number will be discussed in more detail in a later section of this chapter.

Ordering is one of the fundamental concepts of number. Until the children can order objects proficiently, they cannot possibly understand the structure of number. Any attempt to teach number to a child who cannot order objects proficiently will result in the child's attempting to memorize the concept rather than to understand it.

Children's early notions of ordering usually result in the use of the comparative words in a crude fashion. For example, words such as "long" and "short" may be interchanged with words such as "tall" or "small." "Large" may be confused with "many" and "small" with "few." The children's use of the words will gradually become more refined, as they do more ordering activities and hear the teacher use the words in their correct context.

Children's first attempts at ordering objects often result in correct responses for the first and last objects in the sequence, but incorrect responses for objects in the middle of the sequence. This is because the position of the first and last objects are determined by objects on only one side of them, whereas the position of other objects is determined by objects on two sides. For example, consider the sequence of Cuisenaire rods.

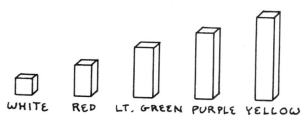

WHITE   RED   LT. GREEN   PURPLE   YELLOW

FIGURE 2–7

The positions of the white and yellow rods are determined only by the red and purple rods respectively (see Figure 2–7). The position of the green rod, however, is determined by both the red and purple rods. Likewise, the positions of the red and purple rods are determined by the white and green rods and the green and yellow rods respectively.

When children begin to work with numbers, it will be necessary for them to order the numbers by looking at both the number that precedes the given number and the number that succeeds it. Therefore, it is clear that children must be able to order objects very proficiently before they attempt to order the more abstract set of numbers.

When preparing ordering activities for children, the teacher should remember to use relationships that are not reflexive and not symmetric, but that are transitive.

Ordering activities are of four basic types:

1. Children may be given a set of objects that are already ordered and be asked to make a copy of the given order. This is the simplest type of ordering activity and should be among the first experiences for children. This type of activity can be made somewhat more difficult by giving the children a sequence and asking them to copy the sequence in reverse order.

2. Children may be given an ordered set of objects and asked to extend the order. The order given should be extended far enough so that its pattern is discernible. The order may or may not have a distinct end point.

3. Children may be given an order that has gaps or missing items in it and be asked to fill in the missing parts. The given order should be complete enough so that the pattern is discernible.

4. Children may be given a set of material having ordinal properties and be asked to invent an order for the set. The materials may or may not have an obvious property with which it can be ordered. The difficulty of the activity can be varied by altering the property by which the material can be ordered.

Materials may be ordered by some qualitative property such as color or shape. Such orders, if they are to be extended or filled in, must utilize some sort of repeated pattern to establish the structure for the order. Orders may also be structured by some quantitative property such as length or size. Such orders are excellent models of the concept of number and are easy to fill in, extend, and invent.

The following are ordering activities that illustrate each of the four categories mentioned above.

1. Lay out a set of four to six objects in a line. Give students an identical set of objects and ask them to make a line just like yours. Use a variety of materials, making sure that you and the students have an identical set. The difficulty of the activity may be varied by asking students to reverse the order

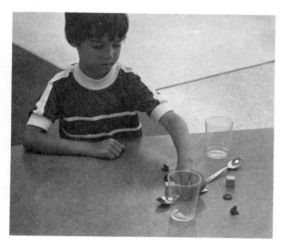

Photo 2-7

of their objects or by laying their objects in a circle rather than a straight line. (See Photo 2-7.)

2.   Select five or six students from the class and ask them to order themselves in some way. Let the other students in the class try to guess how they have ordered themselves.

3.   Give students a set of soda bottles that have varying amounts of colored water in them. Ask the students to order them any way they like. (See Photo 2-8.)

Photo 2-8

4. Give students a set of small pieces of sand paper having various degrees of coarseness. Place the first three pieces in order from coarse to fine and ask the students to place the rest of the pieces in their correct place.

5. Give children a set of common household items that vary in degree of hardness; for example, fork, sponge, and plastic bottle. Place the first three items in order from hard to soft and ask the children to place the remaining items.

6. Give children a set of clear plastic containers and a pouring medium, such as water or sand. Ask the children to put water or sand in each container so the first container has the most, the second a little less, and so on, until the last has none at all.

Photo 2–9

7. Take a comic strip from the Sunday newspaper and cut it up, mounting each frame or group of two or three frames on oaktag. Give the frames to children and ask them to put them in order. Try to choose a comic strip with an appropriate reading level or one with no reading at all.

8. Choose two fixed objects in the classroom that are close to the same length. The width of the teacher's desk and the width of a classroom window will usually suffice. Give children pieces of string and ask them to determine which of the two objects is longer. Use suggestive questions to encourage them to compare the string to both objects. This is an illustration of the transitive property. That is, the desk is longer than the string, and the string is longer than the window, so the desk is longer than the window.

These activities are examples of the numerous ordering experiences available for children to investigate. Children should become so experienced with ordering activities that they can order any set of objects that the teacher chooses to give them. This ability is an absolute prerequisite to an understanding of number.

Equivalence Relations

Equivalence relations possess all three of the properties mentioned earlier. Consider the relation "is the same shape as" on the set of attribute blocks. Investigation will show that this relation is reflexive, symmetric, and transitive. Thus, it is known as an equivalence relation.

To determine if the relation is an equivalence relation, one first asks, "Is each object in the set related to itself?" Because each object in the set is the same shape as itself, the relation is reflexive. Second, if attribute block A and attribute block B are members of the set of attribute blocks, and if A is the same shape as B, then will B be the same shape as A? Because the answer is yes, the relation is symmetric. Third, if A, B, and C are from the set of attribute blocks and if A is the same shape as B and B is the same shape as C, then is A the same shape as C? Because the answer is yes, the relation is transitive. Because the relation is reflexive, symmetric, and transitive, it is an equivalence relation.

When organizing a set by classification, it is often desirable to develop a classification system that has the following two properties:

1. Every member of the set should be included in one of the classes of the system. That is, no member of the set should remain unclassified.
2. The classes in the classification system should not overlap. That is, no member of the set should fit into more than one of the classes.

An equivalence relation is a very precise way to define for a given set a classification system possessing these two properties. If an equivalence relation is given for a set, the relation separates the set into subsets known as *equivalence classes.* The subsets determined by the equivalence relation are such that every single member of the set belongs to one of the subsets, but no member belongs to more than one of the subsets. Thus, a perfect classification system is formed where each object is classified into some subset, but the subsets do not overlap.

If children use the relation "is the same shape as," to associate pairs of attribute blocks, they will discover that all of the triangles are related to one another, all of the circles are related to one another, and so on for squares, rectangles, and so forth. If all of the blocks that belong together are grouped by the "is the same shape as" relation, the blocks will be classified as shown in Figure 2–8, p. 54.

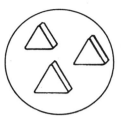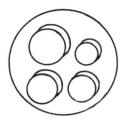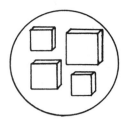

FIGURE 2–8

Each block in the set will fit into one of the subsets, but no block will fit into more than one set. Thus, an equivalence relation on a set produces a very precise classification structure for the set.

Children should experience numerous examples of equivalence relations even though they may never learn the name of the structure they are experiencing. Equivalence relation activities for children are very similar in appearance and structure to the classification activities given earlier in this chapter. There is, however, a very important difference. Classification activities may include almost any sorting activity in which objects are grouped by their properties. Equivalence relation activities should be carefully organized so that every member of the set of objects fits into a class and the classes do not overlap.

On pp. 38–39, a set of classification activities were given. Activities 1, 4, 5, and 6 are equivalence relation activities using the relation, "is the same color as," and relations indicating having or not having wheels, familiarity with the name of a picture, and whether or not an object floats. Activities 2 and 3 may not be equivalence relation activities depending on the objects chosen for the set. These activities used the relations "belongs in the same room as" and "is made of the same material as." For example, in activity 2, if a towel were in the set, it might fit into both the "kitchen class" and the "bathroom class" by the "belongs in the same room" relation. This causes overlap between two classes and violates one of the properties of equivalence relations. The violated property is the transitive property of equivalence relations. For example, the fork "belongs in the same room as" (kitchen) the towel, the towel "belongs in the same room as" (bathroom) the comb, but the fork does not belong in the same room as the comb. Thus, the relation is not transitive, even though it is reflexive and symmetric.

Activity 3 on page 38, in which the child is asked to sort junk into two piles, plastic or metal, would not illustrate an equivalence relation if the set contained an object made of leather since the leather object would not fit into either of the classes given.

Examples of equivalence relation activities are as follows:

1. Give the children pictures of dogs, cats, cows, gerbils, birds, and other animals. Provide two to five different examples of each type of animal. Ask the children to sort the pictures by using the relation "is the same type of animal as."

2. Give the children an assortment of objects that may be grouped into classes of the same length. For example, a spoon, a paperback book, and a pencil might be one of the classes. Provide enough material for four or five classes having five or six objects the same length. Be certain that the objects intended to be the same length are nearly exactly the same and that the objects intended to be in other classes are different by at least two or three centimeters. Ask the children to sort the objects by using the relation "is the same length as."

3. Give the children a set of attribute blocks and ask them to sort them using the relation "is the same color as." There are several other equivalence relations on the set of attribute blocks since they are specially designed to illustrate the equivalence relation concept.

4. Have the children group classmates by the relation "is the same sex as."

5. A more difficult activity, but one that illustrates the reflexive property of equivalence relations, is to have the children group themselves by the relation "is in the same family as." In most classrooms, this relation will put each child in a class by himself or herself, illustrating that each child "is in the same family as" himself—or as the reflexive property states, each member of the set is related to itself.

In these examples, the children should be encouraged to test each object with each other object to provide experience with the three properties that make the relation an equivalence relation. For example, in activity 2, the children should say: "The spoon 'is the same length as' the book; the spoon 'is the same length as' the pencil; the book 'is the same length as' the spoon; the book 'is the same length as the pencil,'" and so on. These statements, taken in a specific order, illustrate the symmetric and transitive properties of equivalence relations. The statements that illustrate the reflexive property, such as the pencil "is the same length as" the pencil, although true and obvious, tend to be confusing to a child. The needed experience with the reflexive property could better be done with activities such as the relation given in activity 5.

You have probably noticed that equivalence relations statements all seem to be of the same form. They usually say "is the same——as" where some property of the members of the set is filled into the blank. This, of course, reveals the nature of equivalence relations and provides a hint of how they will be used later on to define equivalence as a relation between sets.

The Mathematical Basis for Exploring Relationships among Sets

Of the many relationships that children may explore among sets the most important is number. Number is the relationship that will eventually be found most useful in everyday life and will suffuse every other mathematical activity in which the children will engage.

Each topic discussed earlier in this chapter is an absolute prerequisite to number. Number is a complex structure of equivalence and order relations. An understanding of number may not be assumed simply because a child can count.

When the children are thoroughly proficient in the skill of ordering and classifying objects, they are ready to attempt the more difficult activity of ordering and classifying sets. But first, a definition of what might be called a "set of sets" is needed.

Consider, for example, a set of color cubes (see Figure 2–9). Within the set of color cubes is the set of red cubes, the set of green cubes, the set of blue cubes, and so on. Thus, the set of color cubes is really a set of sets.

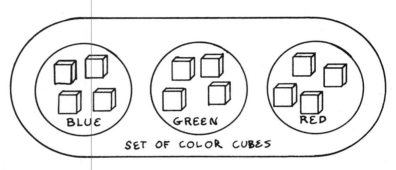

FIGURE 2–9. Set of Color Cubes

When an equivalence or order relation is defined that relates one set to another set, then the sets so related are each one element of a set of sets. So the relation is defined on a set as before, but now the set is a collection of sets rather than a collection of objects.

Equivalence Relations on Sets

One of the most powerful relations between sets is the one-to-one relation. A somewhat formal statement of the one-to-one relation is as follows:

The statement that there is a one-to-one relation between sets A and B means that each element in set A can be matched with exactly one element in set B, so that each element of B is used in a match; and each element of set B can be matched with exactly one element in set A, so that each element of set A is used in a match (See Figure 2–10, p. 57.)

In sets A and B, all conditions are met except one. One of the members of set A is matched with more than exactly one member of set B.

In sets C and D, all conditions are met. The sets are said to have a one-to-one relation between them.

In sets E and F, all conditions are met except one. When set E was matched to set F, one of the members of set F was left unmatched.

In sets G and H, all conditions are met except one. When set H was matched to set G, one of the members of set G was left unmatched.

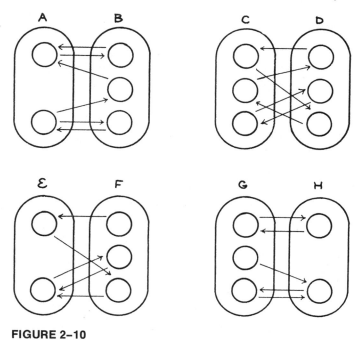

FIGURE 2-10

Even though the children may never see a definition of the one-to-one relationship as detailed as given above, it is important that they have many experiences that illustrate the one-to-one relationship and many experiences in which the relationship is not true as in the examples above.

As the children begin to work with the concept of number as a property of sets, the teacher's knowledge of the children's stage of cognitive development becomes of crucial importance. In particular, the teacher should be aware of whether or not each child has acquired the concept of *conservation of number*. Conservation of number is the ability to understand that the number property of a set does not change when the members of the set are rearranged.

A child who has not acquired conservation of number should not be expected to draw correct conclusions when attempting to match sets. Moreover, he should not be expected to count accurately since counting involves matching a concrete set with an abstract one. The following activities all require conservation of number for successful completion. That children are not "conservers," however, should not eliminate them from participation in the activities. For the "nonconservers," the activities are a means by which the observant teacher can monitor the students' abilities and understand the reasons for any lack of success. Moreover, the children will be in an environment in which they can demonstrate their newly acquired skill the moment that acquisition occurs. The teacher should not allow the students to become discouraged by experiencing too much lack of success. An answer that may be incorrect to an adult, or to anyone who has acquired conservation, may be perfectly correct to a nonconserver. Encourage the children to participate in the activities, but be certain to provide reinforcement for the nonconservers, as well as for those children who give the "adult" answers.

The following activities illustrate the classification aspect of the concept of number:

1. Make a set of eight or nine posterboard triangles in eight or nine different colors. Make a set of circles using the same colors. Present the children with the two sets and ask if they can make a match between them. The children will match the colors, thereby matching the two sets in a one-to-one fashion. This is known as provoked correspondence. The children will actually perform the task by matching colors rather than sets, making this a good introductory activity. The activity would have been more difficult if the sets had each been of a uniform color. When the children have completed the activity successfully one or two times, remove one or more pieces from one of the sets and ask the children to make a match. This allows the children to see an example of two sets that are not related by the one-to-one relation.

In subsequent activities, give the children both one-to-one and not one-to-one experiences. Both are essential to the accurate and thorough acquisition of the concept.

2. Place a set of seven or eight pictures of children so that all can see. Ask one of the children to give each pictured child a piece of candy. (Provide about ten pieces of candy.) If the children succeed with the task, rearrange the pieces of candy used in the match into a compact pile and ask the children if there is still a match between the sets. "Is there still one piece of candy for each child?" Ask one child to prove her answer by reestablishing the original one-to-one match. This activity may also be classified as provoked correspondence.

3. Collect a set of fourteen empty pint milk cartons. Using plaster of paris or playdough, fill seven of the cartons so that one weighs 50 g, another 100 g, and so on in 50 g increments up to 350 g. Paint each of the cartons a uniform color. With the remaining seven cartons, make an identical set of weighted boxes but paint them a different color. Give the sets of boxes to the children and ask them to make a one-to-one match among them. When the children have tried the activity a few times, remove one of the boxes from one of the sets and let the children try to make a match.

If this activity is too difficult for the children, reduce the number of boxes in each set to three or four. This activity may also be classified as provoked correspondence.

Photo 2–10

4.   Give the children sets of color cubes that vary in number; for example, 3 green, 5 blue, 3 red, 2 black, 4 white, and so forth. Ask the children to find the sets that have a one-to-one relation between them. Vary the sizes of the sets used, occasionally making three or four of the sets the same size. If difficulty is encountered when the set sizes are larger than five, the teacher should suspect lack of acquisition of conservation of number. The activity may be simplified by allowing the students to test the equivalence of the sets by stacking the blocks of one set directly on top of the blocks of another.

5.   Give the children a set of six to eight cups and a set of ten to twelve sticks (see Figure 2–11). Ask the children to put a stick in each cup. Then take the sticks and lay them end to end in front of the row of cups. Ask: "Are there as many sticks as cups?" On this activity, the children who are nonconservers will often indicate that there are more sticks than cups because the line of sticks is longer than the line of cups. This indicates that the child is unable to conserve number.

6.   Provide the children with 30 to 40 cards, on which are shown various groups of objects ranging in size from one to ten objects. Thus, there will be three or four cards representing the same number of objects. Also supply about 25 pieces of yarn about 30 cm long each, having a small washer attached

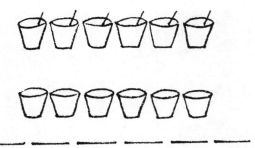

FIGURE 2–11

to each end. Ask the children to place the cards face up and to show matchings between the sets using the pieces of yarn and connecting all sets that can be matched. The children should be able to sort the entire set of cards into ten distinct groupings.

7. Make up a set of number demonstration bars from one to ten as shown in Figure 2–12.

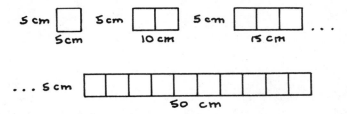

FIGURE 2–12

Provide the children with a set of counters (such as beans or buttons) that are individually much smaller than the 5 cm squares on the number demonstration bars. Ask the children to make a set of counters that matches the set of squares on a given number bar. The children may do it by placing one counter inside each square. Remove the matched set of counters and place them in a compact pile. Ask: "Are there still as many counters as there are squares?" Whatever the answer, ask the children to prove it by returning the counters to their place on the squares.

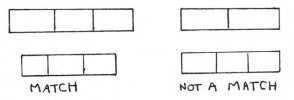

FIGURE 2–13

8. Make a set of number bars having squares 2 cm on a side and another set of bars having squares 3 cm on a side (see Figure 2–13). Ask the children to match the bars that *represent the same number*. The children may have a tendency to match the 2 cm bar representing the number 3 with the 3 cm bar representing the number 2 because they are the same length. Be sure that the children understand that they are to match the bars according to the number they represent rather than the length of the bars.

9. Have the children divide themselves into groups by any variables they choose, such as wears glasses, likes strawberry ice cream, has black hair. Make up 5 to 6 groups with 4 to 6 students per group. Ask the children if any of the groups can be matched.

10. Ask the children to hold up as many fingers as they have eyes, noses, toes, heads, arms, and so forth.

11. Give the children sets of Cuisenaire rods; for example, 3 yellow, 3 green, 3 blue, 4 black, 4 red, 5 dark green, 5 orange, 5 brown, 5 white. Ask the children to find which sets can be matched with one another. This activity may be difficult since the children may be distracted by the lengths of the rods rather than the number of rods per set.

12. Make a chart of posterboard and stand it on a table as shown in Figure 2–14.

Have each child select an inch block and stack it beneath the month in which he or she was born. Remind the children that there is "one block for each child and one child for each block." Have the children gather together in groups according to the month of their birth. Ask the children if there is a match between the groups of children and the groups of blocks. Ask if there is a match between the set of children born in February and the set born in June. Point out that questions such as these can be answered by matching groups of

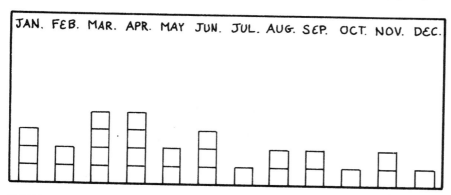

FIGURE 2–14

children or groups of blocks. Ask questions leading to the discovery by the children that they can draw the correct conclusions in this case by looking either at the number of children, the number of blocks, *or the height of the stack of blocks.* This discovery is closely related to the topics of graphing and measurement.

13. Put a set of about nine blocks in front of the children. Using counters such as beans or buttons, ask the children to make a set of counters that will match the set of blocks. Put the set of counters on a piece of paper and ask the children to mark one tally on the paper for each counter. Ask the children if there are as many tallies as blocks. Repeat the activity with a variety of sets of different sizes. This activity is an example of the transitive property of equivalence relations.

Order Relations on Sets

The one-to-one relation on sets is a required prerequisite to comprehending the order relation on sets. The definition of the order relation on sets includes as one of its components the one-to-one relation. The definition of the order relation on sets is: That set A is related to set B by the order relation means there is a one-to-one relationship between set A and some proper subset of set B. (A subset of set B that is not set B, itself.)

In Figure 2–15, there is a one-to-one relation between set A and a proper subset of set B. Notice that set A is also related to set C and set D. However, sets B, C, and D are not related to set A. This implies that the relation is not symmetic. Neither is set A related to itself by this relation, and thus the relation is not reflexive. However, set A is related to set B, and set B is related to set C, so if set A is also related to set C, then the relation will be transitive. And it is. Thus, the relation described above is not reflexive, not symmetric, but it is transitive. This combination of properties describes a relation that is *perfectly ordered.* A perfectly ordered relation on a set assigns each member of the set, on which it is defined, to a specific spot in a sequence. Each member of the set has its unique spot, and no other member of the set shares that position in the sequence. In arithmetic, the relation described above is known as the "is less than" relation. It is defined as a relation between sets, thus the relation is defined upon a set of sets in the same fashion as the one-to-one relation.

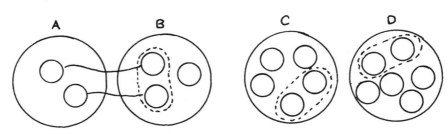

FIGURE 2–15

Because this relation is rather complex, it is important that the children see many varied examples of it. The children also should be very familiar with the one-to-one relation so they will be able to use it confidently when working with the order relation. When children match sets using the order relation they will also be using the one-to-one equivalence relation as a tool. Working with both types of relations at the same time is a complex undertaking and begins to reveal the very sophisticated structure of the number concept, which is incorrectly assumed to be one of the simplest concepts in mathematics. The teacher must be careful to use language that fits with the children's knowledge and experience. In using activities, the language should describe the action to be performed by children rather than the mathematical language.

The following activities require the children to order sets according to the relation described above.

1. Make up ten stiff posterboard cards with sets of household objects glued to them, for example, 1 thread spool, 2 paper clips, 3 sugar cubes, and so on up to 10 items (see Figure 2–16). Provide the children with 10 or 12 pieces of yarn about 30 cm long having small weights on the end. Ask the children to determine which sets can be "matched with only a part of another set." Children will find that the set with only one object in it can be matched with a part of every other set. The set with ten objects cannot be matched with any other. Partial answers should be accepted. If children are encouraged to lay a set "to the left of" any set to which it is related, the final result should be a complete left-to-right ordering of the sets.

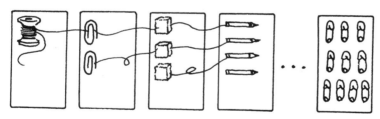

FIGURE 2–16

2. Give the children a set of cardboard cards with dots as shown for the numbers 2 through 9. Also supply a number of map tacks. Ask the children to put map tacks in the loops to represent the set that "comes before" and the set that "comes after" the given set. An alternative activity would be to make up similar cards but give the two outside sets and ask the children to supply the set that "comes between."

3. With a flannel board and several different sets of flannel board objects, put a set of 4 to 7 flannel pieces on the board and ask the children to make a set that "comes before" and a set that "comes after" the given set (see Figure 2–17). If the given set has 6 objects in it, any set 1–5 is correct for

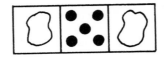

FIGURE 2–17

"comes before" and any set 7–10 is correct for "comes after." Eventually, the children should be encouraged to make the set that "comes *just* before" the given set.

4.   Provide the children with several small juice cans and a pile of about 60 tongue depressors or sticks. Make up sets of size three and seven by placing the correct number of sticks in two of the cans. Ask the children to find the sets that "come between" the two given sets (see Figure 2–18). Any answer 4, 5, or 6 is correct, but encourage the children to try to find all of the sets that "come between" the two given sets. Repeat the activity several times using different given sets. After the children can correctly complete the activity, give them the sets 6 and 7 and ask them to explain why no set can come between them.

FIGURE 2–18

5.   Provide the children with sets (1 to 10) of small objects sealed inside jars. Ask the children to order them from left to right depending upon whether a set can be matched one-to-one with a proper subset of the set on its right. This is a much more difficult activity since the matching cannot be done by physically manipulating the sets. Augment this activity by occasionally leaving one or more jars out of the sequence to allow the children to find the gap. Also, add an extra jar so that there are two jars having the same number of objects. Let the children discover that two sets may fit into the same spot in the sequence.

6.   Provide the children with an egg carton and a supply of beans or other small counters. Place a single bean in the lower left-hand compartment of the carton and ask the children to place the "next set" in the compartment to the right. If necessary, remind the children that they need "one more than" each time to continue the sequence.

Photo 2–11

7. Have the children divide themselves into groups by any random criterion; for example, brown hair, shoes that tie, and so on. If a child fits into two groups, let the child choose the one preferred. Ask the children to order themselves according to the size of their groups. You may wish to pass a strong cord around the outside of each group to keep the group together.

Discuss with them the fact that two groups may fit in the same spot in the sequence, if they can be matched one-to-one. Let children who fit into more than one group change groups, so that the groups will need to be reordered.

8. Give the children cards having various numbers (0 to 7) of objects depicted on them; for example, one card may have 3 dogs, another 3 baseball bats, and another 5 skateboards. Include four or five examples of each set size. Ask the children to classify the cards into groups. After they have completed the classification task (using the one-to-one relation), ask them to order the groups of cards. The final result should be similar to Figure 2–19.

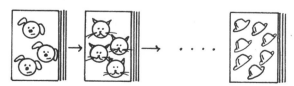

FIGURE 2–19

These activities use concrete or semiconcrete objects that are easy for the children to match one-to-one. After children become confident with both equivalence and ordering activities, they can begin to use semiabstract materials to do their activities. The experience with semiabstract activities will get them ready for the introduction of the number symbols which are completely abstract.

Some examples of semiabstract activities using tally marks follow.

1. Allow the children to keep score in classroom activities where score is kept by a simple tally system. For example, divide the class into teams, alternately allowing children to toss a bean bag into a loop. Assign a box to each team and drop a tally into the box for each bean bag in the loop. After several tosses, let the children take the tallies from the boxes and compare them.

2. Have the children divide themselves into groups by using easy properties, such as "wears a belt or does not wear a belt." Let the children march past a count keeper, who makes one mark on the board for each child in the group. After both groups have been tallied, let the children use another piece of different colored chalk to match the tally marks one-to-one by drawing a line between them. Let the children decide which group is larger, or if they are the same size.

Photo 2–12

Concept of Number

Before it can be assumed that children are capable of understanding number, the children must be able to perform matching or equivalence relations on concrete examples of sets for sets of size 1 through 20. Numerous trials must be given to each child to assure that the child is not fooled by the redistribution of the members of a set of any size. Moreover, the children should be able to take any set of sets from size 1 through 20 and order them by using the one-to-one relationship. Counting may or may not be useful to children attempting this task since most children, even though they can call the number names, will not realize that the statement "17" represents a set larger than the statement "15."

The following activity is a culminating activity, which brings together the concepts of the one-to-one relation and the order relation to illustrate the set theoretic concept of number. Even though only one activity is given, it should be varied using many different types of materials to be certain that the children will acquire a comprehensive abstraction of the concept of number.

1. Make up about 40 cards with pictures cut from magazines representing the sets from size 1 through 6; for example, four people in a group, two automobiles, one house. Include several examples of each of the sets. Place several large loops of yarn on the floor and put one example of each set in each loop. Allow the children to place the other cards into the correct loops so that each loop contains a set of equivalent sets. Paper plates may be substituted for the loops to facilitate moving the groups around.

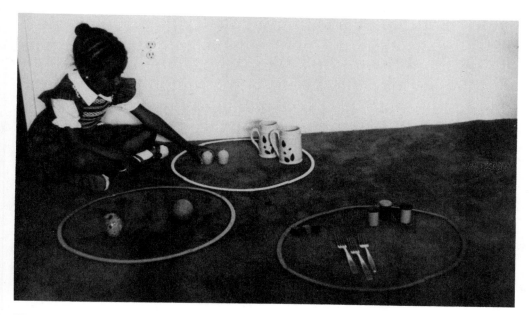

Photo 2–13

Ask the children if they can move the loops of sets so that they are "in order." If the children order the loops according to the number of cards in the set, do the activity again, but arrange the cards so that there will be an equal number of sets in each loop. Suggest to the children that there is still a way to order the loops. Persist until the children discover that they can order the loops by "sets of ones," "sets of twos," and so forth, although the children may not use the number names.

This is a rather complex activity, but it is an accurate depiction of the concept of number. Number is an ordered set of equivalence classes. The equivalence classes are generated by the one-to-one relation, and the equivalence classes are ordered by the "is less than" relation.

After children can confidently manipulate, sort, and order sets by this combination of relations, they may be taught the abstract labels for the ordered set of equivalence classes—one, two, three, four, ...

Extending Yourself

1. Design and construct three concrete activities for children that embody a transitive (order) relation.

2. Design and construct three concrete activities for children that embody the concept of classification.

3. Design and construct three concrete activities for children that embody the principle of the one-to-one relation.

4. Write a two-page report on "Some Basic Processes Involved in Mathematics Learning" by Zoltan P. Dienes in Schminke, C. W. and Arnold, William R. *Mathematics Is A Verb*, The Dryden Press, Inc., Hinsdale, Illinois, 1971.

5. From one of the following references, read about the conservation tasks related to number and length. Prepare the materials needed and try some of the tasks with a five- or six-year-old child. Using a small cassette recorder, record your interview with the child. Write a report of the results, including direct quotes from the child.

Bibliography

Baratta-Lorton, Mary. *Workjobs*. Reading, Massachusetts: Addison-Wesley, 1972.

Copeland, Richard W. *How Children Learn Mathematics*. New York: Macmillan and Co., 1974.

_____ . *How Children Learn Mathematics: Teaching Implications of Piaget Research*. New York: Macmillan and Co., 1970.

Furth, Hans G. and Wachs, Harry. *Thinking Goes to School: Piaget's Theory in Practice*. New York: Oxford University Press, 1974.

Minnesota Mathematics and Science Teaching Project. Units 1, 2, 3, 4, 8. Minneapolis, Minnesota: University of Minnesota, 1969.

Nuffield Mathematics Project. *Beginnings*. New York: John Wiley and Sons, 1972.

_____ . *Shape and Size 2*. New York: John Wiley and Sons, 1972.

_____ . *Shape and Size 3*. New York: John Wiley and Sons, 1972.

Philips, John L. *The Origins of Intellect: Piaget's Theory*. San Francisco: W. H. Freeman and Company, 1969.

C H A P T E R

How Children Attach Meaning to Number

Children Encountering Numbers

Little meaning is attached to the sounds as young children count. To the child they are just sounds chanted in order. Teaching the meaning of number is the challenge of the responsible elementary teacher. Laying the foundation for number is the challenge of the responsible primary teacher. Asked how many buttons are displayed, a four-year-old most likely will begin to count. As the child counts, much may be learned about his understanding of number. Many children count by reciting number names in the correct order but rarely understand what they mean. Here are some examples that typify a child's lack of understanding:

1. If there are six buttons on the table the child will count from one to ten before his finger reaches the last button.
2. If buttons are arranged in a circular pattern or randomly displayed the child becomes confused while counting and must begin again.
3. Asked to put three buttons on a dish the child may put the third button on the dish.
4. If buttons are placed in a pile the child thinks there are fewer buttons than if they are spread out.

It is important for teachers to be aware of and cope with these misconceptions. In some cases coping may be waiting until the child matures sufficiently.

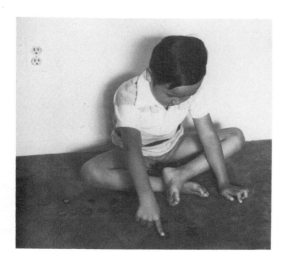

Photo 3-1

Preschoolers typically conserve small numbers of objects. To conserve means the child confidently declares there are five objects regardless of their configuration.

Most three- to five-year-olds need work with numbers up to five. The teacher's job is to assure that the child has a good understanding of these numbers before attempting to teach larger numbers. It is desirable for children to attach the abstract concept of number to concrete materials. Children need a variety of concrete experiences, before they can begin to abstract the notions of number.

The child begins to realize that common experiences in the environment can be easily copied and effortlessly discussed. He begins to connect his tactile experiences with his verbal experiences. One way to describe a set is by using a number to tell how many objects there are. The same number may be used to describe every set with as many objects. By matching any given group of objects and a "standard set," the child is able to determine the number of objects in the given set. More importantly, the child begins to see that "three" is a word referring to all groups of objects that can be matched with a standard set such as that in Figure 3-1. If the child is unable to make this step in understanding, he should still have the ability to make groupings of three objects by matching.

FIGURE 3-1

An Initial Understanding of Number

Children sometimes have difficulty identifying number as a property of sets. The number property of a set may not be as readily seen as properties of the objects themselves, such as color or shape. Try giving a three-year-old a group of cards such as those shown in Figure 3-2.

FIGURE 3–2

Ask the child what is alike about the cards. Responses will vary. Soon the child should note that each card has as many objects as the others. This response should be reinforced with the idea of "twoness," and if the child has not mentioned it, the term "two" should be introduced.

Other activities used to develop the concept of "two" are listed below.

1.   Have children hold up their hands. Note that they each have two hands. If they pick up an object in each hand, how many objects will they be holding? See how many groups of two around the room they can find. Each time discuss how many objects are contained in a group. When a child picks up more or less than two objects, discuss how they can determine whether they have two objects. Use the idea of matching with the child's own hands.

2.   Ask the children to find parts of the body that come in twos. Let them draw their answers. We may expect eyes, ears, feet, hands, legs, and arms. Do not be surprised if elbows, thumbs, and big toes are drawn. Creative children may come up with others. Encourage those unique responses.

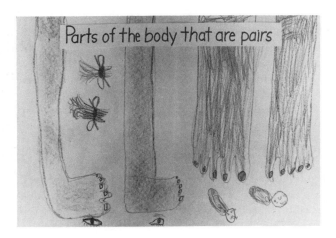

Photo 3–2

FIGURE 3–3

Ask the children how they could show that they have two of each of those parts. The children might draw lines matching each set to a given set, or they may put hands on each part to show they have a part for each hand with no parts left over.

3.   Give the child a box of toys. Pick out a doll and a block. Ask the child to make as many groups that have the same amount as your group as he can. How many toys are in each of these groups?

4.   Make groups such as those shown in Figure 3–3. Ask the child which set does not belong. Why not? Be aware that children may pick sets other than the set with three objects and have a valid reason. Do not discourage this. Accept any reasonable response and continue until the child responds that all sets have two objects but one.

5.   Give the child a set of cards with the pictures as shown in Figure 3–4.

FIGURE 3–4

Ask the child to choose which of the cards shown in Figure 3–5 goes in the given set. Ask him why he chose that card.

FIGURE 3–5

Notice that all these activities develop the concept of number using "two" as an example. When the child has mastered this concept, introduce the numbers one and three. These activities may be adapted to teach one and three, and similar activities may be invented that are appropriate for the children. Numerals are not yet introduced. The numbers are discussed orally with the children. Counting is not used. Children learn to associate number with a set by simply looking at the set.

Try some of the following activities, after the children can identify sets of one, two, and three.

1.   To play "Magic Number" a handful of blocks or buttons and a cover are needed. Hide several blocks under the paper. Lift the paper briefly enough for the child to see the buttons or blocks but not to count them. Let the child tell the "magic number." If he is correct, try another number. If he is incorrect, show him the blocks again for a slightly longer period of time. Children may also play this with each other. A child who has a good concept of the numbers one through three may play with a child who still needs work.

2.   Give a child a set of dot cards with one, two, or three dots. Let him sort them into three piles according to the number of dots on them. Let the child tell how many dots are on each card. He may sort them something like the arrangement shown in Figure 3–6.

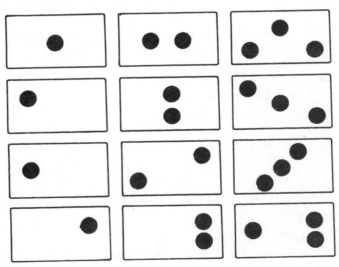

FIGURE 3–6

When the child grasps the meaning of the numbers one through three, introduce four and five using similar activities. Make sure the child can correctly associate numbers through five with sets without counting. Youngsters should be able to look at a set and tell how many objects it has. The position of the objects should not affect the child's response.

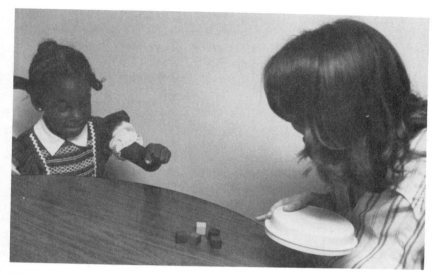

Photo 3–3

Children Encountering Numerals

Once the children can associate the numbers one through five orally with objects, they are ready to learn the numerals. Introduce only one or two numerals at a time. Let the children become familiar with the numerals before asking them to write numerals. Numerals are not something children discover on their own. They will need to be told that the symbol for one is 1, the symbol for two is 2, and so forth. Give the children several opportunities to match numerals with sets.

The following are activities that should help the child learn to recognize numerals from one through five.

1. Make a set of puzzles from tagboard or tri-wall for matching the numerals from one to five with dots or pictures. The pieces might look like those in Figure 3–7. Puzzles of any type are good for self-checking activities.

FIGURE 3–7

2. Cut a picture from a magazine or coloring book the size of a box lid. Mount the picture on tri-wall or posterboard, and cut it into puzzle pieces with a jigsaw (or scissors). Draw around the pieces in the proper position on the box lid. On each section of the box lid, write a numeral. On the corresponding puzzle piece put that number of dots. The child must match the numeral to the dots to make the puzzle fit.

3. A Bingo game will help the child learn colors as well as numbers. A typical card might look like the card in Figure 3–8.

Blue	Green	Red	Yellow	Black
1	4	3	5	2
4	5	1	1	3
3	1	Free	4	1
2	2	4	2	5
5	3	2	3	4

FIGURE 3–8

GREEN

FIGURE 3–9

The caller holds up a color coded dot card such as the dot card in Figure 3–9. The child places three green chips on 3 under green. When a child calls Bingo, the caller confirms that the numerals are covered with the proper number of chips.

4. Several card games reinforce the learning of numerals. Games similar to "Old Maid" or "Fish" afford the child opportunities to match numerals with pictures and with dots. See Figure 3–10.

FIGURE 3–10

5. Dominoes with numerals on one side and dots on the other are useful, as in Figure 3–11. The children are told that dots must be matched to corresponding numerals and numerals to corresponding dots (see Figure 3–12).

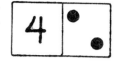

FIGURE 3–11

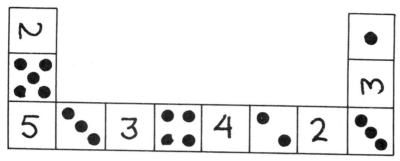

FIGURE 3–12

When children have had experiences recognizing numerals, they begin to write numerals. There are many ways children practice writing besides filling pages with numerals. A few suggestions are given here.

1. On squares of cardboard about 20 cm square, write large numerals and cover them with clear contact paper. Children use a grease pencil to trace over the numerals. They use a cloth to wipe off their work and may repeatedly trace the numerals. When children have traced the numerals several times, they may write the numerals on a sheet of paper.

2. Pour enough salt (or sand) in a box lid to cover the bottom of the lid. The children trace numerals in the salt. They lightly shake the lid to erase the numerals and start over.

Photo 3–4

3.   Cut numerals out of sandpaper and glue them onto tagboard. The children trace the numerals with their fingers. Trace the same numeral in the air and then write it down.

All of these activities should be restricted to the numerals one through five, until the children have mastered them. Mastery includes the children's ability to conserve. Later, the activities are extended to include other numerals, as the children learn them.

Ordering Numbers

Children should be ready to order numbers, when they have learned the meaning of the numbers one through five by using equivalence relations with sets and have had a variety of experiences with order relations. Having ordered objects and sets of objects in many ways, such as length, area, volume, shades of color, and amount, children may extend these relations to number. Thus, the idea of counting is developed meaningfully and not by rote. The children learn that two comes before three, because a set of two has fewer objects than a set of three. Children begin by comparing two sets to find which has more objects. After comparing pairs of sets, the children compare three or more sets by comparing the first two sets, then the third, and so forth. Children should order sets from least to greatest and from greatest to least.

The following activities help children develop the concept of ordering numbers. The symbols for the "greater than" ($>$), "less than" ($<$), and "equals" ($=$) relations may now be introduced.

1.   An adaptation of the card game War is played with a deck of cards having from one to five dots on each card. When constructing the cards, make four cards for each number of dots. The game is played by two children. The cards are shuffled and all are dealt. Both players turn over one card and place them face up side by side. The player with the greater number of dots wins. If the number of dots is equal, "war" is declared. Each player then places one card face down on his first card. He puts another card face up on top of this. The new face up cards are compared. The player whose card has the greater number of dots wins all six cards in the "war." Play continues until one player has all of the cards or a time limit expires. If the children question which group is greater, they should establish a one-to-one correspondence between the dots to decide.

The second phase in this game is to use a regular deck of cards or write the numerals 1 through 5 on 3 × 5 file cards. Again, four cards for each numeral is sufficient. Play continues as before.

2.   Make a puzzle numbered 1 through 5 with a corresponding number of dots beneath the numeral (see Figure 3–13). The pieces are ordered from least to greatest.

FIGURE 3–13

Discuss the idea of a number being one less or one greater than another number.

3. Make a set of dot cards with numerals on one side and dots on the other. Using the dots, the child puts the cards in order from the fewest to the most dots. Turn the cards to the numeral side to check. Reverse the process by ordering the numerals and checking on the dot side using a one-to-one correspondence. Reinforce the relations, "one less" and "one greater."

4. The symbols for greater than (>) and less than (<) may be introduced using "Jaws." "Jaws" eats as much as possible, so his mouth is always open to the greatest amount. Have several pictures or objects available with corresponding numeral cards (see Figure 3–14). The child places the numeral cards under the corresponding sets, decides which is greater (using one-to-one correspondence, if necessary), and places "Jaws" so his mouth is open to the larger numeral. If done on paper, the child may trace inside "Jaws'" mouth to keep a permanent record of the larger number.

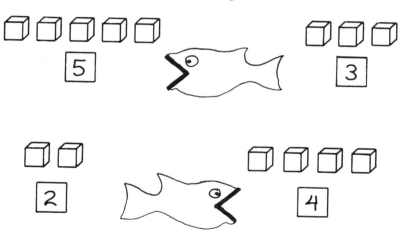

FIGURE 3–14

Discuss the terms "greater than" and "less than," as the symbols are introduced.

The symbol for equals (=) may be introduced at the same time. "Jaws" cannot decide which group to eat, because both are the same size. He keeps his mouth neither wide open nor closed (see Figure 3–15).

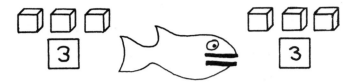

FIGURE 3-15

Most number work for preschool children should emphasize the concept of number and de-emphasize rote counting. A variety of concrete experiences is important. The numbers from one through five should be stressed, with the realization that many young children cannot conserve numbers above five. Larger numbers are introduced as the children demonstrate readiness for them. This may occur well into first grade. Children are introduced to numbers orally in conjunction with objects and pictures; then they learn to recognize numerals and to associate them with groups; and finally they learn to write the numerals. Relationships among numbers are explored, after the meaning of each number is learned. Thus, a sound foundation for learning numbers and their uses in the early school years is developed.

Children Extending Number Concepts

As children extend number skills in their first years at school, they broaden their understanding of number. Concrete materials continue to be an integral part of their learning experiences. Activities to introduce children to number are expanded to help children learn larger numbers. Early concepts are reinforced as new ideas are presented.

In the first two years of school, children often perform in what Piaget terms the transitional stage between preoperational and concrete thought. Conservation is developing in number and measurement. Logical thinking, that is connected to concrete materials, is beginning.

Children are introduced to:

1. The meaning of zero
2. Combining two numbers to make a larger number and partitioning larger numbers into smaller ones
3. Number as a unit of measurement and as a property of sets
4. Number names and their association with sets
5. Grouping objects as a foundation for understanding place value

Activities related to the above concepts follow, both for the children who understand and those having difficulties. All children will not be ready for expanding their number concepts at the same time. They should learn these ideas at their own rate. Pushing an entire class along together may cause confusion and frustration, the faster children being held back or outdistancing the

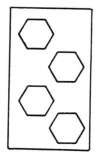 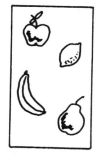 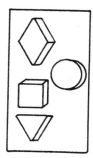

FIGURE 3-16

slower ones. A record of each child's progress would be helpful in planning for each individual. A discussion of ways to specifically cope with children's learning styles may be found in chapter 8.

The Concept of Zero

As children become confident in using the numbers one through five, they are ready for the concept of zero. Zero presents unique problems to the child. A youngster may state that he has no cookies, but it is difficult for him to see that he has *zero* cookies. Having zero seems to be a more difficult idea than not having any.

Discuss what happens if we take away an object from a group that has one object. Act this out with several examples. What number is one less than one? Do we have a number to name this group? Many early numeration systems had no name and no symbol for zero. The invention of zero and its symbol was a significant advance in early systems.

Some children may ask, "What number is one less than zero?" It might be mentioned that we do have negative numbers that are less than zero, but it will probably be advantageous to discuss integers later. By all means carry the discussion to the limits of the children's questions.

Activities useful in developing the concept of zero are now presented.

1.   Have the children name things of which there are zero in the classroom. Among the answers may be "pink elephants," "monsters," and "ghosts," or more practical items, such as "a rug to sit on," "time to play with clay," and "group projects." After each suggestion, discuss whether there are any in the classroom. Let the children draw pictures of something of which they have zero. What about the child who does not draw anything? A blank page is a good picture of zero objects.

2.   Prepare several cards, each with a distinct number of objects pictured. For example, have three cards each picturing the number properties one through five. The "four" cards might look like those shown in Figure 3-16.

As well, there should be three blank cards. These latter cards represent the empty set with the number property zero.

Mix the cards and have individuals sort them in their own way. The object is to see whether or not the children are able to perceive zero in the same way as they perceive groups of one, two, three, four, and five.

3. Collect several small boxes. Put pennies in every box except one. Leave that box empty. Tape each box shut and ask the children if they can find the box with zero pennies without opening them. What would happen if you tried this with pieces of paper, grains of salt, or cotton balls?

4. Put beads of different sizes, shapes, and colors in a box. Describe a certain bead, and ask the children to find all beads of that type and tell you how many there are of that type. Be sure to include descriptions of some beads that are not in the box.

5. When attendance is taken, discuss how many children are absent. How many are absent if you have perfect attendance? What will happen if zero children enter the room? What if zero children leave? Can you hear zero children singing? Ask zero children to stand up, talk, and clap their hands.

Extending a Child's Number Concepts Through Nine

Use the activities discussed earlier in this chapter to extend number concepts from zero through nine. Be sure the children understand each new number introduced. The written numeral may be introduced with each new number. The numbers six through nine may present some new difficulties. Children can visually recognize up to five objects, but may be unable to recognize more than five objects. They may use *rote counting, counting on,* or *grouping* to find the number of a larger set. In rote counting, the child begins with one and counts until he reaches the last object. To count on, the child begins with a number of objects he can recognize on sight, either four or five, and then counts until he reaches the last object. In grouping, the child breaks a larger number of objects into two or more smaller groups, which he can recognize, and combines these to find the total number of objects.

The following activities are suggested to help the child move from rote counting to counting on to grouping.

1. Give the child six buttons. Ask him to separate them into groups, in as many ways as he can, and record the results. Increase the number to seven, eight, or nine buttons. How can the child recognize this many buttons without counting?

2. Take three buttons and then four. How many buttons do we have, if we put them together? Can we get seven, starting with any other groups? Try this with other numbers.

3. Get a set of dominoes with up to five dots on one side. Have children sort the dominoes according to the total number of dots. There might be piles such as shown in Figure 3–17.

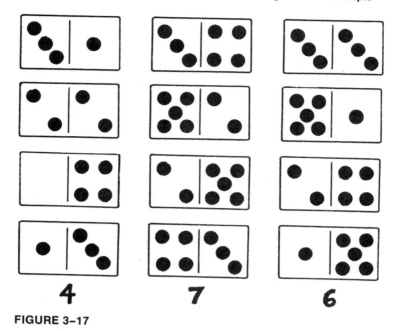

FIGURE 3–17

4. Put a partition in a small box with a lid, as shown in Figure 3–18.

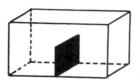

FIGURE 3–18

Place six to nine beans in the box. Put the lid on and shake the box. Take the lid off and record the position of the beans. If seven beans are used, the result might be 6 and 1, 4 and 3, 2 and 5, 0 and 7, and so forth. Try this with other numbers of beans, shaking the box several times for each group of beans.

The Number Line

Thus far the activities have focused on the number of objects in a grouping. Number is also used in measurement. The number line and ruler are common examples. Measurement is employed when we say, "John lives *three* blocks away," "Julia is *four* years old," "It is *two* o'clock," and "Mr. Smith drives *eighty* kilometers an hour." Measurement is considered in detail in chapter 6. The number line, however, extends the concept of number, and it will be discussed briefly.

The number line is based on units of length. "Two" is more than just the number marking the point halfway between one and three. Two describes the distance from the starting point, or zero, to the point marked two, based on an arbitrary unit of measure. This can be a difficult concept for young children for two reasons. First, children are accustomed to associating numbers with groups of discrete objects. The number line is based on length, rather than on a number of objects. Second, many youngsters between five and seven years old are unable to conserve length. It is difficult for them to understand that the distance from 0 to 2 on a number line is twice as long as the distance from 0 to 1.

Following are activities to introduce children to the number line. A full understanding of the number line will develop through time and use, as the child develops in the understanding of conservation of length and the concept of measurement.

1.   Make a large walk-on number line on the floor with masking tape. Note that the starting point is marked 0. No steps have been taken at this point. Do not begin a number line with one, since this detracts from the length concept. Mark on the number line intervals, which are the size of a child's step (about 25 cm). Make sure each interval is the same size. The children take turns starting at zero and walking a given number of steps. Have them confirm that the point at which they finish corresponds to the number of steps they took. In this way the number of discrete steps is associated with the distance walked on the number line.

Photo 3–5

Photo 3–6

2.   From oaktag, construct desk-sized number lines for each child. Mark on them the numerals 0 to 10. Laminate or cover them with clear contact paper, so that the children may write on them with grease pencils. Ask children to show various points on the number line, by beginning at zero and counting the appropriate number of spaces. What number is on the point where they land? Emphasize counting spaces rather than points.

3.   Francine Frog is helpful on the number line. Use the desk number line, and let Francine help. If you want to show "3" on the number line, Francine starts at zero, hops three spaces, and lands on 3. This reinforces the idea of counting spaces rather than points.

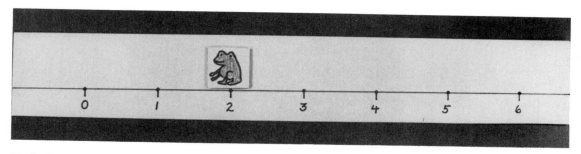

Photo 3–7

Once the child has a good concept of the numbers from 0 to 9 and can recognize and write the numerals from 0 to 9, he is probably ready to learn to read and write the written number names. This may be done at any time the child encounters these words. The activities that were developed to teach the number line may be extended to include number names. Here are a few other ideas.

4.   There are several good children's books, which emphasize learning number names. Many of these associate number names with numerals and pictures. While reading these stories, let the children find the numerals and check to see if the pictures show the correct number of objects.

5.   Construct a "rummy" game in which the child needs a numeral card, a number name card, and a picture card to make a set. This idea may also be used for other card games such as Fish or Old Maid. Sets may look something like Figure 3–19.

FIGURE 3–19

6.   Let the children attach number names to their number lines. This may be done to the walk-on number line as well as the individual desk-sized number lines.

Photo 3–8

An Introduction to Grouping

Grouping is important to our system of numeration. It would be difficult to symbolize and manipulate large numbers without some type of grouping. For this reason, it is important for children to begin grouping objects as they are developing the concept of the numbers zero through nine. Because place value is of such great importance to our numeration system, and because grouping is fundamental to place value, extensive grouping work is essential for young children. It is also a basis for understanding fractions. Therefore, children should have experience grouping by 2s, 3s, 4s—all the way to 10.

Grouping materials may be proportional or nonproportional. Proportional materials are constructed so that if groupings are by 10, the material for 10 is ten times as large as one, the material for 100 is ten times as large as ten, the material for 1000 is ten times as large as one hundred, and so forth. Nonproportional aids do not show this consistent size distinction (see below). Proportional materials include Multibase Arithmetic Blocks, tongue depressors, bean sticks, and Cuisenaire cubes, squares, and rods (see Figure 3–20). Nonproportional aids include chip trading, the bottle game, money, and the abacus (see Figure 3–21).

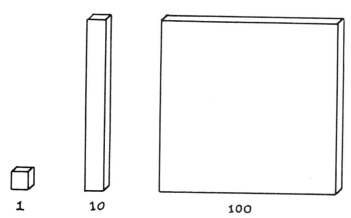

1   **10**   **100**

FIGURE 3–20. Proportional Grouping Material

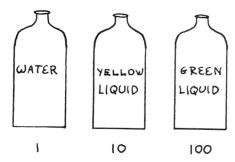

FIGURE 3–21. Nonproportional Grouping Material

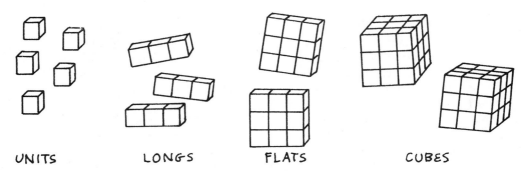

UNITS   LONGS   FLATS   CUBES

FIGURE 3–22

The following activities are designed to give children thorough experiences in grouping. This becomes the foundation for later work with place value.

1. Multibase Arithmetic Blocks (M.A.B.) are sound proportional materials for grouping. They consist of various sizes of wooden blocks representing powers of particular grouping points (size of each grouping being used). The M.A.B. are commercially available in sets with grouping points of 2, 3, 4, 5, 6, and 10. The set of blocks with a grouping point of 3 is shown in Figure 3–22.

WOULD BE EXCHANGED FOR

FIGURE 3–23a

FIGURE 3–23b

FIGURE 3–24a

FIGURE 3–24b

The first activity with the M.A.B. after initial free play is discovering the grouping point and establishing exchanges. Once it has been decided that three is the grouping point, children should be given handfuls of units and asked to make all exchanges possible. For example, the blocks in Figure 3–23a would be exchanged for the blocks in Figure 3–23b because the blocks in Figure 3–24a represent the same amount of material as the blocks in Figure 3–24b. There are two such units left over. If children have difficulty making these exchanges, they should be instructed to make groups of three units, until all units are used, and then exchange each group of three for a long.

2. Proportional materials may be constructed by the teacher from railroad board or other firm construction material. Figure 3–25 illustrates one such set with a grouping point of three and Figure 3–26 illustrates one set with a grouping point of four.

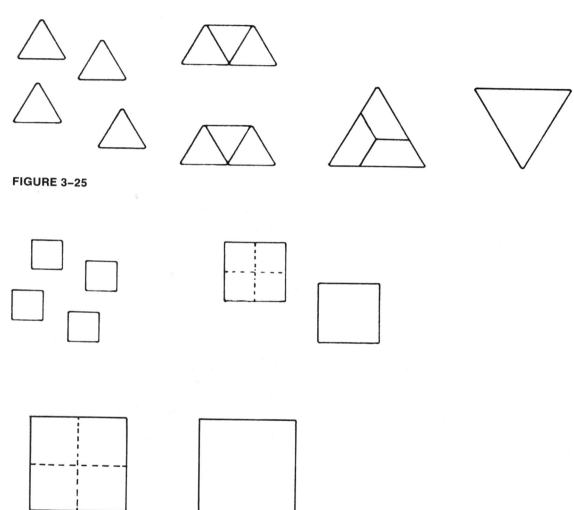

FIGURE 3–25

FIGURE 3–26

Each activity suggested for the M.A.B. may be played with the materials in Figures 3–25 and 3–26.

3. Another proportional material that the teacher may construct is the bean stick. For a grouping point of five there would be loose beans, sticks with five beans attached representing a first grouping, five sticks of five attached representing a second grouping, and so forth (see Figure 3–27).

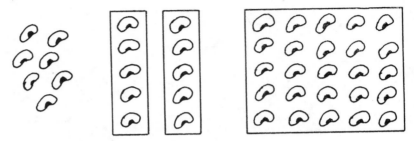

FIGURE 3–27

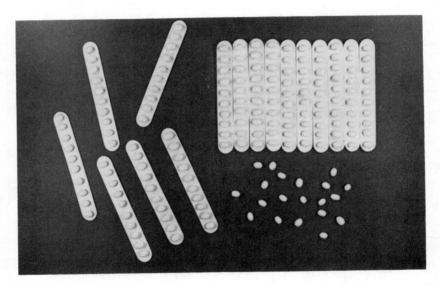

Photo 3–9

Again, each of the activities described may be used with the bean sticks. Notice that each of the above proportional materials were shown using one or two grouping points. The teacher is encouraged to have the children perform groupings and exchanges with several grouping points. A single type proportional material may be used for all groupings.

4.   Begin with 8 buttons. Have the child place these in groups of *** and record his results on a chart as in Figure 3–28.

GROUPS OF ***	ONES
2	2

FIGURE 3–28

Make groups as indicated in Figure 3–29, and record the results.

GROUPS OF ❋ ❋ ❋ ❋	ONES

GROUPS OF ❋ ❋ ❋ ❋ ❋	ONES

GROUPS OF ❋ ❋ ❋ ❋ ❋ ❋	ONES

GROUPS OF ❋ ❋ ❋ ❋ ❋ ❋ ❋	ONES

FIGURE 3–29

Groups of * * * *	Ones

Groups of * * *	Ones

Groups of *** ***	Ones

Groups of *** * ***	Ones

FIGURE 3–30

The second phase of this activity begins when a group of 25 buttons is considered. The first step is still to place all objects in groups of ***. Next, three groups of *** are grouped together. It is helpful to label the first groups of three, "groups," the second grouping, "super groups," and so forth. Thus, we have the configuration as in Figure 3–31.

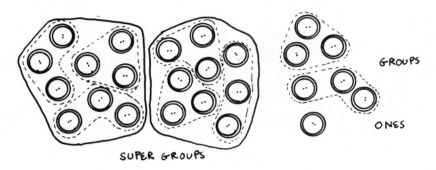

FIGURE 3–31

The record for this expanded grouping would be with 3 tally boxes, as in Figure 3–32. Continue this activity with grouping by four, five, six, and so on. Record the results. Also, try starting with other quantities of buttons.

SUPER GROUPS	GROUPS	ONES
2	2	1

FIGURE 3–32

5. The Bottle Game. Collect about 100 empty baby food jars with screw-on caps. (Hospitals will often give you theirs.) Fill the jars with water, and use food coloring to tint them different colors. You should have about 40 clear, 30 yellow, 20 green, 20 red, and 10 blue. Choose a magic number for trading. The children should take turns rolling a die and collecting that many clear bottles. When the magic number is reached, exchange clear bottles for a yellow. For example, if the magic number is 3, 3 clear bottles may be traded for one yellow, 3 yellows for one green, 3 greens for one red, and 3 reds for one blue. Set the goal as either a red or a blue.

The bottles may be used for all the same games as the bean sticks and the multibase blocks. Children are very motivated by the trades.

Photo 3-10

6. Colored chips are convenient nonproportional teaching aids. They may be purchased commercially or constructed by the teacher. Any value for the chips may be established in the same manner as in the bottle game. For example, 4 yellow are equivalent to 1 blue, 4 blue are equivalent to 1 green, and 4 green are equivalent to 1 red. The activities used in the bottle game are appropriate for the colored chips.

Photo 3-11

Other grouping activities are presented in the next section on numeration systems. Activities at this point should be informal and should use simple recording systems. When the child formally encounters place value, he should be familiar with grouping from these and similar activities.

Grouping work should be extensive and children should be comfortable with groupings of various sizes. Grouping sizes increase until they reach ten. Because of the importance of ten to the Hindu-Arabic system of numeration, grouping by ten becomes a major focus. When grouping by ten is performed, it is understood because of the students' previous work. Likewise, the concepts of the basic operations are presented, using objects in various sized groups with varying grouping points.

Extending Number Concepts through Twenty

The teacher should find the following activities useful in extending the understanding of numbers between ten and twenty. Place value is presented as a natural outgrowth of the grouping experiences.

1.   The activities used to develop the numbers from 6 to 9 should be repeated in developing the concept of 10. Children should be familiar with a variety of ways to make 10.

2.   Take out the orange and white Cuisenaire rods. Begin with 10 white rods. Ask the child to find a rod for which these could be traded. How do you know that 10 whites are the same as one orange? Try this with 11 to 15 white rods. Can you trade for more than one orange? How many whites are left?

Photo 3–12

3. Get beads of the same size, and cut string long enough to hold exactly 10 beads. Give the children between 10 and 20 beads to string, and let them fill out the following chart.

Full Strings	Beads Left	Total Number of Beads
1	4	14
1	2	12
1	0	10
etc.		

FIGURE 3–33

Fractions at the Intuitive Level

In their first years of school, children have an intuitive grasp of the fraction concept. One half means to the child that something has been separated into two parts, regardless of the size of each part. "Your half is bigger than mine," is a frequent comment. Children at this stage are ready for a simple introduction to the meaning of halves, thirds, and fourths. Separation of objects into equivalent parts should be stressed. Introductory activities for fraction work are suggested.

1. Have several rectangular pieces of oaktag, each of which is separated into two pieces in different ways (see Figure 3–34). Two children are to share a

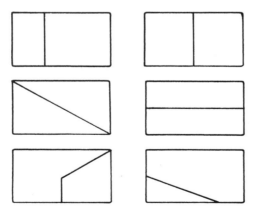

FIGURE 3–34

rectangular shape so each will get the same amount. Let the children manipulate the oaktag to show their solutions. Indicate that these pieces are each half.

Have the children compare those chosen as halves and those not chosen. Repeat the activity with thirds and fourths.

2. Fold each of several sheets of paper into two parts. Let the children identify which sheets are folded in half. Give the children paper to fold into halves. Let them check each other. Are any papers folded into two unequal parts? Repeat this with thirds and fourths.

3. Draw pictures of several familiar objects (see Figure 3–35). Separate them into two pieces in several ways. Let the children find which objects are separated into halves. If the child identifies the halves, give him pictures of ob-

FIGURE 3–35

jects which are not divided. Let him show halves. This may be repeated later for thirds and fourths.

Activities for further development of fraction and decimal concepts are presented at the end of this chapter.

Children Symbolizing Number

Once the child has learned the symbols for the numbers from 0 to 9, there are no new symbols to learn. These ten symbols are used to name all the natural numbers in our numeration system. The child now needs to learn how and why these symbols are put together to form larger numerals. Place value is integral to understanding our base ten numeration system. Grouping concepts, which were introduced in the last section, will serve as the foundation for understanding place value. The development of concepts and symbols for common fractions will also build on the grouping idea.

Seven-to-nine-year-old children often perform at what Piaget terms the "concrete operational" stage. They have skills that the preoperational child does not have. They conserve number and length. They use multiple classification skills and form series of objects in a variety of ways. These and other skills help the child learn more complex ideas than have previously been possible. There are, however, some limitations to the concrete operational child's skills. Much understanding comes from manipulation of concrete materials. The child has difficulty thinking abstractly about a problem. This should be kept in mind as concrete tasks are developed.

Concepts at this stage are built on earlier ideas. New concepts about number and numeration emphasized at this stage are:

1. The meaning of place value.
2. The meaning of fractions.
3. Fractions and decimal fractions.

The Meaning of Place Value

Place value is a feature of writing numerals that relies on three things: (1) digits, (2) size of the particular groupings used, and (3) the grouping point. The number of digits used is limited by the choice of grouping point. If the grouping point is $\overset{**}{**}$, then any numeral written will be limited to the digits 0, 1, 2, and 3. The power of the grouping indicates how many positions are being used, as well as the relative size of the number considered. Consider the collection of objects in Figure 3–36:

SUPER GROUPS   GROUPS   ONES

FIGURE 3–36

The numeral 213 describes the amount shown when the grouping point of $\overset{**}{**}$ is employed. The place-value grid in Figure 3–37 can be used for recording the number in each group.

SUPER GROUPS	GROUPS	ONES
2	1	3

FIGURE 3–37

The numeral has three digits, all of which are either 0, 1, 2, or 3. The numeral extends to the second power of the grouping point $\overset{**}{**}$.

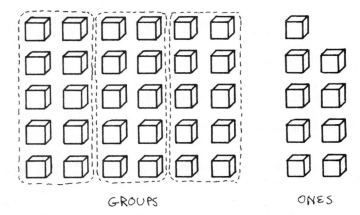

GROUPS   ONES

FIGURE 3–38

With a grouping point of *****, the same number of objects shown in Figure 3–38 would be written 39, meaning three groups of ***** and nine units. The place-value grid in Figure 3–38a shows how the numeral *39* is recorded. With a grouping point of *****, the digits 0, 1, 2, 3, 4, 5, 6, 7, 8, and 9 may be used. When working with children, the teacher should help them understand the pattern as one moves to the left or right on the place-value grid. Understanding that each position to the left is ten times the value of the position to the immediate right and that each position to the right is one-tenth the value of the position to the left is of utmost importance.

Super Groups	Groups	Ones
	3	9

FIGURE 3–38a

Children need many concrete activities to form a good basis for understanding place value. The activities described earlier will provide most of this understanding. The place value concept should be extended after children have a firm foundation based on concrete materials. Here are a few activities for abstract reinforcement to be used after understanding is achieved.

1. Make three sets of cards numbered from 0 to 9. On one set of the cards write ones, on the second set write tens, on the third set write hundreds. Pass out the cards to the class. Call out various numbers. If 267 is called, children with these cards would get together the array in Figure 3–39. Continue until every child has had a chance to participate.

2 hundreds	6 tens	7 ones

FIGURE 3–39

2. Make a deck of cards with matches such as:

600	6 hundreds	six hundred
50	5 tens	fifty
4	4 ones	four, etc.

Use the cards to play Rummy, Fish, and Old Maid.

3. Make fold-out cards that tell place value, such as in Figure 3–40.

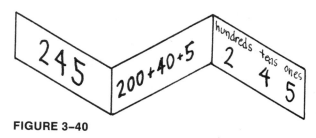

FIGURE 3–40

These activities deal with decimal numeration. Once understanding of grouping and place value is achieved, the child should be encouraged to calculate with decimal numbers. During or after the above activities, it would be appropriate to discuss the ideas of place value: digits, groupings, and grouping points.

The Meaning of Fractions

The seven-year-old should be able to converse, using the simple fractional terms halves, thirds, and fourths. He should differentiate between an object broken or cut into two, three, or four pieces, and the same object cut into halves, thirds, or fourths. That each third of an object is the same size is basic to understanding fractions. The fractional concept is now expanded to include the idea of a unit amount. An object, set of objects, or drawing is declared as having the value of one unit or one. If it is then separated into three equivalent parts, each part is one third of the object.

Some children have difficulty comparing a part of a unit to the unit itself. They may not have reached the Piagetian level of class inclusion. Instead of comparing the part to the unit, they compare one part to the other parts. For example, a child may call the shaded sections in Figure 3–41 "one half." This child is comparing the one shaded section to the two nonshaded sections. This

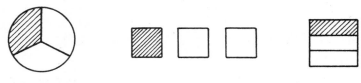

FIGURE 3–41

ratio idea of comparing two parts is not generally introduced until a later stage. Activities at this level should include a variety of materials.

1. Early work with the Cuisenaire rods will likely have developed ideas such as, "Red is one half of purple, because it takes two reds to make a purple." Familiarity with fractions will grow from exploring fractional relationships demonstrated by the rods. Let brown have the value of two. Can you find the rod having the value of one? If purple is one, what value does red have? Because it takes two red rods to make a purple rod, each red rod is one-half of purple. Suppose the red rod has the value of one. Can you find a rod with the value one-half? Because it takes two white rods to make a red, each white rod is one-half of red. White is one half of red, and red is one-half of purple. Can you find other pairs of rods to show one-half? Draw pictures of them. An informal discussion would likely center on why rods such as yellow and blue do not have corresponding rods worth one-half.

Fractional relationships of one-fourth and one-third emerge from exploring length relationships in a manner similar to the above. Various rods are assigned the value one. Thus, green is one-third of blue, because it takes three greens to make a blue. Reinforce that each green rod is the same length.

Photo 3–13

Photo 3–14

Once the concepts of one-half, one-third, and one-fourth have been developed with objects and drawings, the numerals should be introduced to show how one-half is symbolized. The symbol ½ has meaning to the extent that the concept has been carefully developed.

Larger fractional amounts with thirds and fourths are an outgrowth of earlier work. If one green represents one-third when blue is one, two greens represent two-thirds, and three greens represent three-thirds. Likewise, if brown is one, red is one-fourth, then two reds are two-fourths, and so on.

2. Fractional ideas developed with the Cuisenaire rods are similarly developed using Pattern Blocks (see Appendix). If the parallelogram has the value of one, then the triangle is one-half, because two triangles make a parallelogram. Three triangles make a trapezoid. Three parallelograms make a hexagon. Two trapezoids make a hexagon. Six triangles make a hexagon. Many fractional relationships are discovered by manipulating and drawing the Pattern Blocks. Children should be guided through a sequence, ending with symbolizing the fractional concepts learned.

3. Figures created on the geoboard (see Appendix) lend themselves well to exploring fractions. How many ways can you use a rubber band to separate the geoboard in half? One group of children was able to show more than 200 different ways to separate the geoboard in half. See how many ways you can separate the geoboard into fourths. Copy Figure 3–42 on your geoboard. Make a figure one-half the size. Make a figure one-third the size. Make three other figures with which you can show one-fourth.

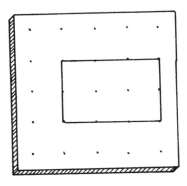

FIGURE 3–42

FIGURE 3–43

4. Prepare a page of square regions with which the children can work. Ask them to separate each region in half in a different way. Have the children compare their results. Discuss why some illustrations do and do not show one half. Discuss why each of the shaded areas in Figure 3–43 represents one half.

Must halves always look alike? Cut the pieces to show the shaded regions are equivalent in area to the unshaded regions. Reinforce the notion that the regions must be equivalent in area. Extend this activity to include thirds and fourths. Later, introduce the fractional symbols.

5. Empty egg cartons and beans or buttons make handy representations of fractional parts. Challenge the children to put beans in one-half of the cups. See how many different arrangements of one-half can be shown. What does one-half mean, when we use twelve cups to represent one unit? Show how one-third can be displayed in several ways. How would you show two-thirds? How would you show three-thirds? In how many ways can one-fourth be displayed? Again, as the children develop facility with the meaning of fractions, encourage them to symbolize their ideas.

6. Prepare sets of tangram pieces (see Appendix), or have the children assist you. A pattern is shown in Figure 3–44. (See p. 103.) Cut the pieces apart.

Explore fractional relationships. How many can be named? For example, each small triangle is one-half of the square, one-half of the medium triangle, and one-half of the parallelogram. The medium triangle is one-half of the large triangle. The square is one-half of the large triangle and equivalent to the parallelogram and the medium triangle. How can these relationships be verified? Try laying the pieces on top of one another. What other relationships can be found? Let the children manipulate the pieces and challenge each other to find fractional parts.

7. Recipes are a practical source of fractions. Have children bring their favorite recipes or provide an inexpensive children's cookbook. The opportunity should be provided for children to measure ingredients and find fractions

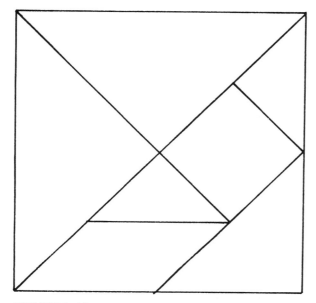

FIGURE 3–44

of recipes. Of course, actually cooking simple foods will help reinforce the applications of fractions. Typical fractional measures are ½ C. of sugar, ¼ tsp. of salt, ½ lb. of butter, ½ C. of butter, and ½ pint of cream. As metric recipes become readily available, they should be included.

Decimal Fractions

Just as the earlier development of the numeration system was built to the grouping point of ten, the development of decimal fractions is built to the fractional parts of tenths. For example, the orange Cuisenaire rod or ten egg carton cups are named the unit or one. Then, a white rod or a single egg cup represents one-tenth. Once activities are presented to highlight tenths, the place value concept with a grouping point of ten is expanded to decimal notation. The following activities help develop the concept of decimal fractions.

1. The activities used with the Cuisenaire rods earlier are extended to include the idea of tenths. The orange rod has the value of one. What value does one white rod have? Because it takes ten white rods to make an orange rod, each white rod is one-tenth of orange. What would be the value of three white rods? What is the value of seven white rods? What is the value of nine white rods? What is the value of five white rods? What is another way to express this amount? How many white rods make a green rod? If the orange rod has the value one, and the white rod is one tenth, what is the value of the green rod? Can you find another way to represent three-tenths?

As the children become comfortable with the concept of tenths and later with the symbols 1/10, 2/10, 3/10, . . . , the teacher should introduce the decimal notation. The value of the green rod, when orange is one and white one-tenth, can be expressed as .3. All rod relationships should be symbolized with the decimal notation. Considerable practice will be needed to master the new notation. Eventually, .1, .2, .3, .4, .5, .6, .7, .8, and .9 will become as easy or easier than common fraction notation.

2. The egg carton can be quickly converted to tenths by cutting two sections from one end. By placing beans or buttons in the remaining ten cups, the concept of tenths can be explored. When seven of the cups have beans, seven-tenths is the focus of attention. Tenths is developed in a manner similar to the other fractional parts discussed earlier with egg cartons. The tenths sequence is similar to that outlined with the Cuisenaire rods above.

3. The metric system of measurement employs decimal notation throughout. Active practice with the metric system will soon have the child needing to use tenths.

a. Find something that is .5 m. How far is it from your desk to the door in meters?
b. Pour 1.4 liters of water into a milk container. Each plant should be given .2 liters of water.
c. Find an object that weighs .3 kg. As children grow in using metrics, so also should they grow in using the decimal notation.

As the measurements become more exact the designations of tenths may well shift to hundredths. The measurement from desk to door may be 2.35 m. Because of the considerable background in developing the concept of place value, the shift to hundredths should not be overly difficult.

4. Money provides a practical example of tenths and hundredths. The values of a penny compared with a dime and a dime compared with a dollar are one-tenth. If a dime has the value of one, four pennies have what value? If a dollar has the value of one, and a dime is one tenth, what is seven-tenths? When we write symbols for amounts of money we include hundredths, so seven-tenths of a dollar is symbolically written $.70 and pronounced seventy cents, where cents is another word for hundredths. A classroom store can offer realistic practice with tenths and hundredths and can prove to be useful outside of school.

5. The base ten pieces from the multibase arithmetic blocks help extend the place-value concept into tenths and hundredths (see Figure 3–45).

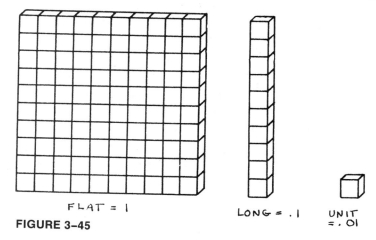

FLAT = 1   LONG = .1   UNIT = .01

FIGURE 3-45

a. If the flat has a value of one, what value will four longs have? Because one long has the value of one-tenth, four longs have the value of four-tenths.

b. If the long has a value of one, what pieces will have the value of seven-tenths?

The relationship can be discovered by manipulating the objects themselves. Development of tenths and later hundredths follows the sequence outlined for the Cuisenaire materials in activity #1 above. Common and decimal fraction notation culminates work with these materials.

To follow up the activities with decimal fractions, the teacher should discuss the pattern discovered earlier in the place-value grid. The pattern that each place is one tenth as large as the place to its left continues, as one moves to the right past the ones-place (see Figure 3-45a). Encourage the children to predict the value of the place immediately to the right of the ones-place. Discuss the use of the decimal point, to indicate the separation between the whole numbers and decimal fractions.

tens	ones
1	1

tenths	hundredths
1	1

FIGURE 3-45a

Extending Yourself

1. Administer Piagetian conservation of number tasks to two children in kindergarten, first, and second grades. Are children in the same grade always on the same level? Do older children always perform at a higher level than younger children? What does your diagnosis indicate about each child's readiness for problems involving number?

2. List pros and cons for teaching rote counting to young children. How might it be detrimental to the learning of number concepts? What are some practical uses of rote counting?

3. Take a position on the following statement: "Young children should not learn place-value concepts in any base other than 10." Support your viewpoint.

4. Make up an original idea for a place-value activity. Be sure to include your objectives. Construct the activity, and try it with one child or a small group of children. Evaluate your idea.

5. Examine two or three textbook series to determine when and how number concepts are first introduced. Are both rote and rational counting used? What consideration is given to the concepts of greater than, less than, and one-to-one correspondence? Is any provision made for children who do not conserve number? When are fractions and decimals introduced?

6. Consider how the same material can be used to teach whole number and fractional number concepts. List three materials that could be used for each, and develop two lessons for each material—one for whole numbers and one for fractions.

Bibliography

Baratta-Lorton, Mary. *Workjobs.* Reading, Massachusetts: Addison-Wesley, 1972.

——————. *Mathematics Their Way.* Reading, Massachusetts: Addison-Wesley, 1976.

Davidson, Patricia S.; Galton, Grace K.; and Fair, Arlene W. *Chip Trading Activities.* Arvada, Colorado: Scott Scientific, 1972.

Dienes, Zoltan P. and Golding, E. W. *Modern Mathematics for Young Children.* New York: Herder and Herder, 1970.

Gibb, E. Glenadine and Castaneda, Alberta. "Experiences for Young Children." *Mathematics Learning in Early Childhood,* Thirty-seventh Yearbook. Reston, Virginia: National Council of Teachers of Mathematics, 1975.

——————. *Teaching Arithmetic Concepts to Pre-primary Children.* Glenview, Illinois: Scott, Foresman and Co., 1969.

Ginsburg, Herbert: *Children's Arithmetic: The Learning Process.* New York: D. Van Nostrand Co., 1977.

Piaget, Jean. *The Child's Conception of Number.* New York: W. W. Norton and Co., 1965.

Stern, Catherine. *Children Discover Arithmetic.* New York: Harper and Brothers, 1949.

C H A P T E R

As Children Begin Number Operations

Children's Concepts of an Operation

Early childhood mathematics typically focuses on the four basic operations with whole numbers: addition, subtraction, multiplication, and division. Developing the concept of *an operation* should begin before the child encounters number work. The child should experience operations in working with objects and sets.

An *operation* on a set is a relation that associates an ordered pair of objects or abstractions from the set with a single member of the set. For example, the operation of addition on the set of whole numbers associates the pair (3, 5) with the number 8. Multiplication associates the pair (2, 3) with the number 6.

Young children should have much experience physically manipulating sets and discovering properties of set operations. This should build on previous work with structure and relations of sets. It should build the foundation for later operations with numbers.

Operations with Objects and Sets

Two operations young children should experience with sets are union ($\cup$) and intersection ($\cap$). It is more important for children to know the concepts of the operations than it is for them to know the formal definitions of union and intersection. Use the children's language to introduce the concepts.

To find *the union of two sets* means to find the set containing everything that belongs to either set. The members belonging to both sets are included, but not repeated, in the union. For example, the union of children with blue jeans and children with sneakers would be all children wearing either jeans or sneakers. Notice this union also includes children wearing both jeans *and* sneakers.

The following activities are designed to help children develop the concept of union. Neither the term nor the symbol are introduced at this point.

1. When selecting children to work in various centers, use the union concept. For example, anyone wearing a blue shirt or red pants may paint. Everyone with black socks or blonde hair may play with the blocks.

2. Use the union idea when playing with building blocks. Build a tower with all the blocks that are red or yellow. Make a house from blocks that are triangular or rectangular.

3. Simple Venn diagrams may be introduced at this point. Colored string or plastic hoops may be used to define the sets. Begin with two sets not having common members (disjoint sets). For instance, you may use buttons in sets that are visualized by string loops (see Figure 4–1).

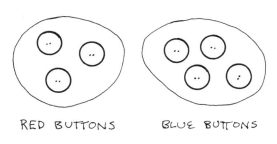

RED BUTTONS   BLUE BUTTONS

FIGURE 4–1

Ask the children to place all the buttons that belong in each set within the appropriate loop. Then ask for the buttons that are red or the buttons that are blue.

Try similar activities using different types of materials and two or more disjoint sets. Sets that overlap may be introduced later, when the child has the idea of intersection.

To find the *intersection of two sets* means to find the set that contains the members common to both of the given sets. This may be a difficult concept for young children in Piaget's preoperational stage. They may not understand that one object may belong to more than one set. For example, you may give children a set of attribute blocks and ask them to put all the red blocks in one set and all the circles in the other. Some children will be unable to decide what to do with the red circular blocks. Some will not use the red circles at all. Some might suggest putting half the red circles in one set and half in the other. A few children might suggest the red circles should be shared between the two

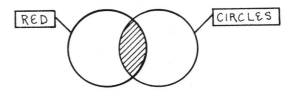

FIGURE 4–2

sets. These children are probably ready to work with Venn diagrams, where the circles overlap (see Figure 4–2). The intersection is where the red circles belong. If children do not understand this concept, they may not be ready to continue with the intersection activities that follow.

1.   Ask the children wearing a blue shirt *and* red sneakers to pass out the milk. Be sure the children who stand up have both a blue shirt and red sneakers, not just one or the other.

2.   Ask all children wearing blue jeans to go to the paint corner and all children with brown hair to go the block corner. Discuss what happens to brown-haired children wearing blue jeans.

3.   See if students can arrange the hoops to show all of the white buttons and all the round buttons (see Figure 4–3).

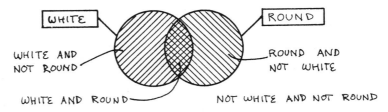

FIGURE 4–3

Can they place all the buttons correctly? Try this with several types of materials.

The *complement of a set* is everything in the universe that does not belong in a given set. For example, in discussing all children in a class, the complement of the set of girls includes all boys in the class. Children who can identify the members of a set sometimes have more difficulty in identifying what is not in the set. The following activities should help children to strengthen this concept.

1.   Ask children who are not wearing red to get out the clay. Ask everyone who does not have a dog to stand in line. See if children respond correctly.

2.   Ask children to put all the red blocks inside the hoop (see Figure 4–4). Discuss the blocks that do not belong in the hoop.

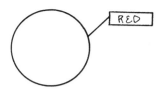

FIGURE 4-4

After children have experienced union, intersection, and complement of sets, see if they are able to use all three ideas.

1.   Have a scavenger hunt. Ask the children to find:

Something not red
Something large and round
Something wooden or green

2.   Play a Simon-Says type game. Distinguish among children by using phrases such as:

"Everyone wearing green or red stand up."
"Everyone not wearing blue jeans clap your hands."
"Everyone with blonde hair and sandals jump up and down."

Properties of Operations

Once children understand the concepts of union and intersection, they may be introduced to the commutative, associative, and identity properties of union and intersection. Again, terms and formal definitions are not as important as understanding the basic concepts. These properties are important to the operations of addition and multiplication. Thus, young children are building a foundation with sets for their later learning about numbers.

When the outcome of union or intersection of two sets is unaffected by the order of the two sets, the operations are said to possess the *commutative property.* Symbolically that may be written:

A ∪ B = B ∪ A
A ∩ B = B ∩ A

Let the children discover this for themselves. Have children wearing red or brown stand up. Then have children wearing brown or red stand up. Are the same children standing? The following activities expand on the idea of the commutative property.

1.   Use Venn diagrams to place blocks such as in Figure 4-5, p. 112.

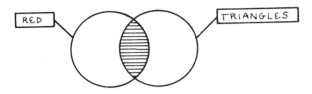

FIGURE 4-5

Have the child find the pieces that are red and triangles. Let the child walk to the other side of the arrangement to get a new perspective. What are the pieces that are triangles and red? Are they the same?

2.   Discuss everyday events to see if the order makes a difference. Is putting on a sock, then a shoe, the same as putting on a shoe, then a sock? Is putting on the left shoe, then the right shoe, the same as putting on the right shoe, then the left shoe?

Try other events such as: eating a cookie and ice cream, putting on a shirt and pants, or putting on a shirt and coat. Let the children suggest others of their own.

3.   Have children find things that are blue or wooden. When they have found several, ask for things that are wooden or blue. Do the children realize that both requests are the same?

The commutative property will be very useful when the children later discover $3 + 4 = 4 + 3$ and $5 \times 2 = 2 \times 5$.

The *associative property* is concerned with three or more sets. At any given time, union and intersection are performed on only two sets. When three or more sets are being considered, the operation is performed on two of the sets, followed by the operation performed on the third set, and so on. It does not

Photo 4-1

matter which two sets are operated on first; the results remain the same. This characteristic of the operations of union and intersection is the *associative property*. Symbolically, the associative property for each operation may be

$$(A \cup B) \cup C = A \cup (B \cup C)$$
$$(A \cap B) \cap C = A \cap (B \cap C)$$

Notice that the order of the sets stays the same. It is only the grouping that changes.

Many young children will have difficulty with this property because they are used to working with only two sets. Be sure children have a firm understanding of operations with two sets before introducing a third.

Some of the following activities may challenge the more advanced students.

1. Use the attribute blocks and a Venn diagram to work a problem such as (yellow ∩ small) ∩ circles.

First work yellow ∩ small, as in Figure 4–6a, and then (yellow and small) ∩ circles, (Figure 4–6b).

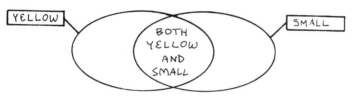

FIGURE 4–6a

FIGURE 4–6b

Work several examples of this type using two distinct steps until the children can do it proficiently. Then have the children reverse the steps in the exercise above by combining the circles and the small blocks first and then combining the result (circles ∩ small) with yellow.

Several more examples will allow the children to discover that (yellow ∩ small) ∩ circles gives the same result as yellow ∩ (small ∩ circles).

2. Use a Venn diagram with three hoops. See if children can place attribute blocks in a diagram such as in Figure 4–7, p. 114.

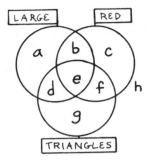

FIGURE 4-7

If children are frustrated with placing pieces in this diagram, they may be guided to skip this section or discuss only disjoint sets. Ask children to find the regions where the large, red blocks may be found (regions b and e). Then ask which of these regions also contains triangular blocks (region e). Now start with the pieces that are red and triangles (regions e and f). Which of these two regions also contains large blocks? (region e). This shows (large ∩ red) ∩ triangles = large ∩ (red ∩ triangles). Try this with a variety of examples using either union or intersection.

The associative property will be used later when children discover that:

(2 + 3) + 4 = 2 + (3 + 4) or
(5 × 2) × 3 = 5 × (2 × 3).

Photo 4-2

Another useful concept is the identity element. The identity concept is used in addition when children learn that any number plus 0 is the same number. Zero is the identity element for addition. In multiplication any number times 1 is the same number results in one being the identity element for multiplication. The identity element for union is a set that, when taken in union with another set, does not change the original set. After some experimenting, children should be able to discover that the empty set is the identity for union. What happens if all the girls are asked to stand and then all the purple people-eaters are asked to stand? No additional children should stand. Was anything joined to the original set? Let the children suggest similar examples.

Several commercial materials are useful for building set concepts. These include People Pieces, attribute blocks, and logical blocks.

All the work with set operations and their properties should be building a foundation for later number work. Set work does not end when number work begins; it should continue to build as number operations develop.

Children Learning the Basic Number Facts: Understanding and Skill

Once children understand the nature of an operation, they should be ready to understand the concepts of operations with whole numbers. All work with addition, subtraction, multiplication, and division should be firmly based on concrete manipulations. Later, the concepts will be expanded to semiconcrete or pictorial representations and finally to abstract number sentences. It is also assumed that children have a good fundamental understanding of place value. Learning the basic addition facts depends heavily upon place value, since they involve sums as high as 18.

This section is concerned with basic facts. For addition, that means only facts with one-digit addends, that is, $0 + 0 = 0$ to $9 + 9 = 18$. In multiplication, basic facts have one-digit factors. The basic facts for subtraction and division involve the same numbers as the addition and multiplication basic facts. For subtraction, that means all facts from $18 - 9 = 9$ to $0 - 0 = 0$; for division, $81 \div 9 = 9$ to $0 \div 1 = 0$. Problems with larger numbers, especially those that require regrouping, will be considered in the next section.

Children's Concepts of Addition, Subtraction, Multiplication, and Division

Early addition concepts should be based on the manipulation of sets. Let the children actually handle the objects. Use examples as they arise in the classroom. Children need much experience with problems such as: "If I have three green blocks and four red blocks, how many blocks do I have altogether?"

1. Have available all types of simple counters, dried beans, one-inch cubes, bingo chips, and so on. Let children use them to solve problems as they arise. Many children may solve problems by repeatedly counting the objects.

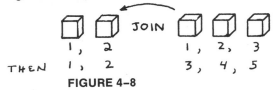

FIGURE 4–8

For example, to add 2 + 3 the child may count out 2 objects, then 3 objects, and then return to the first object to start counting all over (see Figure 4–8).

Encourage children to recognize the number of objects in a set, and use counting as little as possible. This builds on earlier number concepts.

2. Use a pan balance. Show that 2 paper clips and 3 more paper clips balance 5 paper clips on the other side. Thus, the pair (2, 3) is associated with 5, because they balance with 5. Balance 2 one kg weights and 3 one kg weights in one pan with 5 one kg weights in the other.

3. If children can conserve volume, use graduated cylinders. Show that if you pour in 1 cup and 3 more cups, the cylinder will show you have a total of 4 cups (see Figure 4–9).

Photo 4–3

FIGURE 4–9

4. If children can conserve length, try joining color cubes by length. Three color cubes joined to four color cubes are as long as seven color cubes (see Figure 4–10).

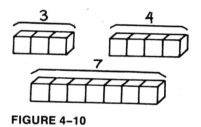

FIGURE 4–10

This is a prelude to the introduction of measurement and the number line concept.

5. Use a walk-on number line made of masking tape on the floor. Show that a jump of 4 spaces and another jump of 2 spaces lands in the same place as a jump of 6 spaces.

6. Children should be familiar with Cuisenaire rods at this point. The rods can be used to reinforce addition concepts. Children try to find single rods which are the same length as a train of two or more rods (see Fig. 4–11).

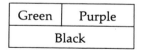

Green	Purple
Black	

FIGURE 4–11

Children must find the black rod to match the train of green and purple. Sometimes whole numbers are associated with the rods. Assuming the white rod has the value one, the green rod will have the value of 3, and the purple rod will have the value of 4. Then green plus purple equals black is equivalent to $3 + 4 = 7$. Children should explore relationships with the rods before introducing the number sentences.

Many children will not be able to understand subtraction as the reverse of addition because they do not yet have the Piagetian concept of reversibility. They cannot reverse 3 + 4 = 7 to get 7 − 4 = 3 or 7 − 3 = 4. They may find it easier to think of subtraction as a "take-away" process. However, this gives a limited view of subtraction. Subtraction can also be used to solve comparison problems of the type: "John is 8, Jack is 6. How much older is John than Jack?"; and missing addend problems of the type 5 + □ = 8. Children might begin by solving take-away problems, but the comparison and missing addend problems should not be ignored.

Many of the same materials used to introduce addition concepts should be used with subtraction.

1. Be sure children have a supply of counters. Try several problems such as: Take five counters. Remove 3. How many are left?

To use the counters for comparison problems, try: "John has 6. Susie has 3. How many more does John have than Susie?" Let the children compare the two sets using one-to-one correspondence to find the solution. In Figure 4–12, John has 3 more.

FIGURE 4–12

In a missing addend problem, ask questions such as: "Jack has 3¢, but he needs 5¢. How much more money does he need?" Let the child use pennies or counters to answer the question (see Figure 4–13).

FIGURE 4–13

This will probably be more difficult than the take-away or comparison problem. Many children in the preoperational stage may not be ready to solve this.

2. The walk-on number line may also be used for these three types of problems:

Take-away: Start at 0. Walk 5 steps forward. Walk back 4 steps. On what numeral do you end?

Comparison: Have 2 children start at 0. One child walks 6 steps. One walks 4. How many steps farther has the first child walked?

Missing Addend: Start at 0. Walk 2 steps. How many more steps will it take to get to 6? Many children who cannot solve missing addend problems using other materials will be able to understand it using the number line.

3.   Cuisenaire rods may also be used in subtraction. They are best for the comparison problem.

How much longer is the dark green rod than the red rod? (Figure 4–14.)

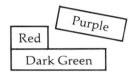

FIGURE 4–14

Children must find the rod they can put next to the red rod to make a train the length of the dark green. Later, as number is applied to the rods, number sentences may be used to describe this as 2 + ☐ = 6 or 6 − 2 = ☐.

Children should experience all types of subtraction using a variety of concrete manipulatives to insure a good understanding of the subtraction concept.

Multiplication should first be introduced as repeated addition. Some children may find multiplication easier than subtraction because it does not require reversibility.

Most American textbooks introduce 2 × 3 as two groups of three, but most British textbooks (and some American ones) will introduce it as two, three times. Do not let children be confused by this. Be consistent with your presentation.

The same materials used for addition and subtraction may be used to teach multiplication.

1.   Use the counters. Stress making groups with the same number in each group. Give each of three children two pieces of candy. How many pieces do you need? Let the children make up problems with counters or candy.

FIGURE 4–15

2.   Use the pan balance. What weight will balance 4 sets of 2 one kg weights?

3.   Use the walk-on number line. On what numeral do you land after 3 jumps of 3 spaces each?

4.   Make a train of 4 red rods. Which rod will match this train? (Figure 4–16.)

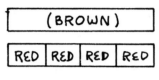

FIGURE 4–16

Number sentences may be introduced later to match the trains.

5.   Rectangular arrays are also useful in showing multiplication facts. This is an arrangement of rows in which each row has the same number of objects. For instance, 2 × 3 may be shown as 2 rows of 3 (see Figure 4–17).

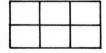

OR

FIGURE 4–17

Multiplication may also be shown as a Cartesian product instead of repeated addition. The Cartesian product matches everything in one set with everything in a second set.

For instance, Jenny has 3 skirts and 2 blouses. If she wears each blouse with each skirt, how many outfits could she make? (See Figure 4–18.)

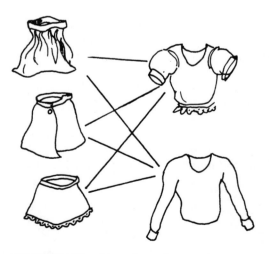

FIGURE 4–18.   Lines show the 6 outfits.

	Blouse 1		Blouse 2	
Skirt 1	S_1	B_1	S_1	B_2
Skirt 2	S_2	B_1	S_2	B_2
Skirt 3	S_3	B_1	S_3	B_2

FIGURE 4–19

Some other examples of uses of Cartesian products follow.

1. How many two-dip ice cream cones can you make if the first dip can be chocolate or vanilla and the second dip may be strawberry, cherry, chocolate chip, or pistachio?

2. How many slates of officers can you make if Kathy, John, and Sally are running for president and Maureen, George, and Sam are running for vice-president?

Let the children make up and solve other examples.

Division problems fall into two categories: measurement problems and partition problems. In a *measurement problem,* the total number of objects are provided along with the number of objects to be put into each group. It is then necessary to find the *number of groups* that can be made. For example, if there

Photo 4–4

Photo 4–5

are 12 pieces of candy and each child is to be given 3 pieces, how many children will receive candy? In a *partition problem,* the total number of objects are provided along with the number of groups that are to be made. It is then necessary to find *how many will go in each group.* For example, "If there are 12 pieces of candy and 3 children, how many pieces are given to each child?" The candy is partitioned into 3 sets.

Children need practical experience with both types of problems. Provide these experiences when it is time to hand out paper or milk; form groups to work on projects; choose teams for games.

The same materials that were used for the other three operations may be used to explore division.

1. Give children a handful of counters. Let them work both measurement and partition problems.

Measurement: Take 15 counters. How many groups of 5 can you make?
Partition: Take 15 counters. Make 5 equal groups. How many are in each group?

2. Cuisenaire rods can also be used for both measurement and partition problems. Children may find the measurement problems a little easier using the rods.

Measurement: How many red rods does it take to make a brown rod? How many yellow rods in an orange?, and so on.

Partition: Find two rods of the same color that are the same length as the dark green. Find three rods of the same color to match the blue.

3.   The number line may also be used for both measurement and partition problems. However, it is easier to use it for measurement problems and is commonly used in that way.

Start at eight on the number line. How many hops of four spaces will it take to return to zero?

Start at twelve. How many hops of three spaces will it take to get to zero?

Extending the Understanding to Semiconcrete and Abstract Levels

Children should have a firm concrete understanding of the operations before moving to the semiconcrete or abstract levels. On the semiconcrete level, pictures are used to represent the actions. In the beginning, number sentences should not be used in isolation. They should be used to record concrete manipulations or to describe an action in a picture. Later, children use the number sentences without either concrete or semiconcrete representations.

Operation and relation symbols should be used only with numerals and not with pictures. It is incorrect to write the problem as in Figure 4–20.

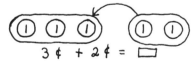

FIGURE 4–20

The following examples in Figures 4–21 and 4–22 show some possible uses of semiconcrete symbolization in addition, subtraction, multiplication, and division.

1.   Addition:

Notice that the total set of 5 is not drawn.

2.   Subtraction:
     Take-away:

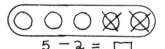

FIGURE 4–21

Comparison:

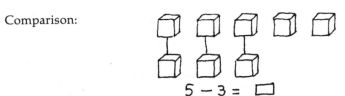

$$5 - 3 = \square$$

Missing addend:

$$3 + \square = 5$$

3. Multiplication:

$$3 \times 2 = \square$$

4. Division:
   Measurement:

$$8 \div 2 = \square$$

Partition:

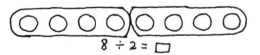

$$8 \div 2 = \square$$

FIGURE 4–21 (*cont.*)

The number line is a good semiconcrete aid for all the operations. The following examples in Figure 4–22 show its use with each of the four operations. Notice the arrows measure spaces or distance, not counting points.

1. Addition

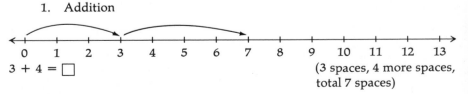

$3 + 4 = \square$

(3 spaces, 4 more spaces, total 7 spaces)

FIGURE 4–22

2. Subtraction
    Take-away:

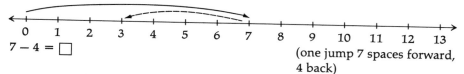

$7 - 4 = \square$ (one jump 7 spaces forward, 4 back)

Comparison:

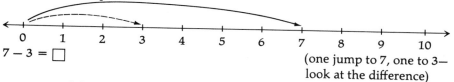

$7 - 3 = \square$ (one jump to 7, one to 3— look at the difference)

Missing Addend:

$3 + \square = 8$ (one jump to 3, how many more spaces to 8?)

3. Multiplication

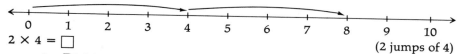

$2 \times 4 = \square$ (2 jumps of 4)

4. Division
    Measurement:

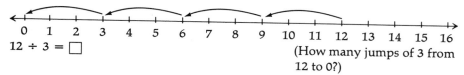

$12 \div 3 = \square$ (How many jumps of 3 from 12 to 0?)

FIGURE 4–22 (cont.)

After children understand the operations for the basic facts on the concrete and semiconcrete levels, they are ready to use the abstract symbols. They should learn to calculate in both horizontal and vertical form. That is, children should be familiar with each of the following forms.

$$\begin{array}{r} 3 \\ +4 \\ \hline \end{array} \qquad 3 + 4 = \square \qquad \begin{array}{r} 9 \\ -2 \\ \hline \end{array} \qquad 9 - 2 = \square$$

$$\begin{array}{r} 3 \\ \times 2 \\ \hline \end{array} \qquad 3 \times 2 = \square \qquad 8 \div 2 = \square \qquad 2\overline{)8}$$

Learning the operations should be connected to problems in everyday life whenever possible. Let children help with collecting milk money and taking attendance. Discuss the mathematics children use at home such as going to the store, getting an allowance, earning money for chores, and saving for a special toy. Use mathematics when problems arise in the study of other subject areas—measuring plant growth, solving problems in stories read, and map making, for example.

As children explore operations with whole numbers, they should discover that the commutative, associative, and identity properties learned with union and intersection also hold for addition and multiplication, but not subtraction and division. These properties are useful as children begin to memorize their basic facts. For example, each time children learn a new addition fact, they automatically learn another, if they know the commutative property. When they learn $6 + 7$ or 8×9, they automatically know $7 + 6$ or 9×8. It cannot be assumed that children know the commutative property. They need to manipulate materials to show $5 + 6 = 6 + 5$ and $7 + 4 = 4 + 7$, but $6 - 4 \neq 4 - 6$ and $6 \div 3 \neq 3 \div 6$. Give children the opportunity to learn and reinforce this property using all the materials used to learn the property itself. Knowing the commutative property will reduce the number of basic facts children must learn for addition and multiplication by almost half.

The associative property is also useful in learning basic addition and multiplication facts. Let children use materials to discover that $(3 + 2) + 4 = 3 + (2 + 4)$ and $(5 \times 3) \times 2 = 5 \times (3 \times 2)$, but $(3 - 2) - 4 \neq 3 - (2 - 4)$ and $(6 \div 2) \div 3 \neq 6 \div (2 \div 3)$. Remember to do the work in the parentheses first.

The usefulness of the associative property is not as obvious as that of the commutative property. It does help children in working with facts they know to build larger facts. For example, suppose children are working $8 + 6$. They know $8 + 2 = 10$. Therefore, they may restate the 6 as $2 + 4$. The problem then becomes $8 + (2 + 4)$. Using the associative property,

$$
\begin{aligned}
8 + (2 + 4) &= (8 + 2) + 4 \\
&= \quad 10 \quad + 4 \\
&= \quad 14
\end{aligned}
$$

Notice that this statement assumes that the children have a thorough understanding of place value. Avoid using this approach with children who do not have a complete command of place value. It will result in severe frustration and discouragement. Children use facts they know to discover new facts.

Here are some other examples:

$$
\begin{aligned}
6 + 9 = (5 + 1) + 9 &= 5 + (1 + 9) \\
&= 5 + \quad 10 \\
&= 15
\end{aligned}
$$
$$
(8 + 3) + 7 = 8 + (3 + 7) = 8 + 10 = 18
$$
$$
6 + (4 + 9) = (6 + 4) + 9 = 10 + 9 = 19
$$

Be careful not to let children get bogged down in symbolism. The property should be used to help children and not confuse them.

The identity elements for addition and multiplication also reduce the number of basic facts children must learn. Once children know that any number plus 0 or times 1 is that same number, they have 10 fewer facts to learn (not even considering the commutative property, which makes it a total of 19 fewer facts). Subtraction and division have a partial identity, called a right-hand identity. This means the identity must be the number on the right. That is: $a - 0 = a$, but $0 - a \neq a$, and $a \div 1 = a$, but $1 \div a \neq a$.

Let children use materials to work several problems with the identity elements. They should be able to discover that $0 + a$, $a + 0$, $a \times 1$, and $1 \times a$ always equal a. Do not assume that children know this automatically, and avoid asking them to memorize it. When they discover it for themselves, they should understand and remember it.

One other property especially useful in learning the multiplication facts is the distributive property. The distributive property helps build larger multiplication facts by adding the sum of two smaller facts. Symbolically it is stated: $a \times (b + c) = (a \times b) + (a \times c)$ or $(a + b) \times c = (a \times c) + (b \times c)$. If children want to compute 5×8, but they only know their multiplication facts to 5×5, they could separate the 8 into $5 + 3$. The problem then becomes:

$$5 \times 8 = 5 \times (5 + 3) = (5 \times 5) + (5 \times 3)$$
$$= \quad 25 \quad + \quad 15$$
$$= 40$$

What would happen if they separated the 8 into $4 + 4$. The problem then becomes:

$$5 \times 8 = 5 \times (4 + 4) = (5 \times 4) + (5 \times 4)$$
$$= \quad 20 \quad + \quad 20$$
$$= 40$$

Try this for other examples. Does it always work?

An array may be drawn to demonstrate the property (Figure 4–23).

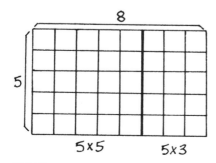

FIGURE 4–23

Figure 4–23 shows the 5 × 8 array broken into the two smaller arrays of 5 × 5 and 5 × 3. Notice it could be split up in several ways: 5 × 4 and 5 × 4; 5 × 2 and 5 × 6; 3 × 8 and 2 × 8; 3 × 5, 3 × 3, 2 × 5, and 2 × 3, and so on.

Many children will be able to discover the distributive property on their own by using arrays such as the one in Figure 4–23. Memorization of the distributive property without understanding often results in confusion and incorrect calculations. Be sure children can demonstrate the property concretely (or semiconcretely) before they are asked to use (or memorize) it.

The following activities are for reinforcement of all four properties. They are to be used after children have discovered and understood the properties.

1.   Make a concentration game with cards that match according to the properties. One set of cards might look like the set in Figure 4–24.

6 + 7	7 + 6
7 + 0	7
9 ÷ 1	9
(6 + 4) + 2	6 + (4 + 2)
(3 + 4) × 4	(3 × 4) + (4 × 4)
8 × (5 + 2)	(8 × 5) + (8 × 2)
(6 × 3) × 2	6 × (3 × 2)
4 × 3	3 × 4
6 − 0	6
4 × 1	4

FIGURE 4–24

Turn all twenty cards over. The first player turns over 2 cards. If they match, he keeps them. If not, he turns the cards over, and it is the next player's turn. The player with the most matches at the end is the winner.

2.   For children who can identify the names of properties, make a set of cards that name a property and a concrete material. The children demonstrate the property using that material. They can check their answers with others in the group or by an answer key. Children take turns drawing cards and demonstrating the property. Some cards might be as in Figure 4-25.

```
┌─────────────────────────────┐
│     Commutative Property     │
│         of Addition          │
│        Number line           │
└─────────────────────────────┘

┌─────────────────────────────┐
│     Distributive Property    │
│       of Multiplication      │
│         over Addition        │
│            Array             │
└─────────────────────────────┘

┌─────────────────────────────┐
│     Associative Property     │
│       of Multiplication      │
│            Blocks            │
└─────────────────────────────┘

┌─────────────────────────────┐
│     Right-hand Identity      │
│        for Subtraction       │
│        Cuisenaire rods       │
└─────────────────────────────┘
```

FIGURE 4-25

3.   For more advanced children, make a Rummy game in which cards match equations with letters rather than numbers. The children try to get a set of three: the name of the property and the two halves of the equation. The set of cards as in Figure 4-26 might be used.

```
┌──────────────────────────┐   ┌──────────┐   ┌──────────┐
│                          │   │  a + b   │   │  b + a   │
│   Commutative Property   │   └──────────┘   └──────────┘
│       of Addition        │
│                          │
└──────────────────────────┘

┌──────────────────────────┐   ┌──────────┐   ┌──────────┐
│                          │   │  a × b   │   │  b × a   │
│   Commutative Property    │   └──────────┘   └──────────┘
│     of Multiplication    │
│                          │
└──────────────────────────┘
```

FIGURE 4-26

| Associative Property of Multiplication | $(a \times b) \times c$ |
| | $a \times (b \times c)$ |

| Associative Property of Addition | $(a + b) + c$ |
| | $a + (b + c)$ |

| Distributive Property of Multiplication over Addition | $a \times (b + c)$ |
| | $(a \times b) + (b \times c)$ |

| Distributive Property of Multiplication over Addition | $(a + b) \times c$ |
| | $(a \times c) + (b \times c)$ |

| Identity for Addition | $a + 0$ | a |
| | $0 + a$ | a |

| Identity for Multiplication | $a \times 1$ | a |
| | $1 \times a$ | a |

| Right-hand Identity for Division | $a \div 1$ | a |

| Right-hand Identity for Subtraction | $a - 0$ | a |

FIGURE 4–26 (*cont.*)

Remember, the concept of the property is more important than the label. Do not rush into memorization of meaningless symbols.

Reinforcing Basic Number Facts

Once children have a good understanding of the operations at the concrete and semiconcrete levels, they should begin to build skill, speed, and accuracy with the basic facts. This requires time and practice. The basic facts should be memorized to provide a foundation for all later computation. Skills can be developed painlessly using a variety of methods once concepts are understood. Children do not have to do a ream of workbook pages to learn facts quickly and accurately.

Children like to follow their own improvement. Let children keep graphs of their progress on timed tests. Once a week, give the children a page of basic facts and three minutes to work them. Let the children check their papers and record how many problems are correct. If the children are motivated to learn the facts, their scores should improve each week. Those children who get all the problems correct may only take the timed test once a month to ensure the continuation of skill. Children may be motivated by becoming members of the "18 Club," when the basic addition facts are memorized, and the "81 Club," when the multiplication facts are memorized.

Children who know their facts well may wish to challenge the teacher. Time the teacher to see how long it takes the teacher to do a page of facts. (Make sure they are all correct.) Children can then attempt to beat the teacher's record.

There are also several commercial and teacher-made games that motivate children to learn basic facts. The purpose of many of these games is to reinforce the facts, not to develop understanding of concepts. They should be used after children know the concepts of the operations.

Teacher-made aids can be very simple. The following ideas may all be constructed from a set of blank 3 × 5 index cards. Give children eleven 3 × 5 cards, and ask them to number them from 0 to 10. The children should keep the cards for all of the following activities.

Many activities may be used for reinforcement of any or all of the basic operations. The teacher may adapt them to suit the children's needs.

1. Bingo. Children should select 9 of their cards and arrange them in a 3 × 3 array, such as in Figure 4–34.

6	2	8
4	1	7
9	5	3

FIGURE 4–27

2	7	6
9	5	1
4	3	8

FIGURE 4–28

The caller calls from a set of problem cards that have 0 to 10 for an answer. The problems may be addition, subtraction, multiplication, or division. Children cover the number that is the answer to the fact. For example, if 3 + 4 is called, children would cover 7. The first player to get three in a row, horizontally, vertically, or diagonally is the winner.

2. Show-Me Cards.   The teacher asks number questions orally, and the children hold up the numeral card with the answer. For example, the teacher may say, "5 + 4 − 3 + 7 − 6," and the children should hold up 7. The teacher can quickly check to see which children can follow the problem and which cannot.

3. Magic Squares and Triangles.   Use the numbers 1 to 9. Put them in a 3 × 3 array so that each row, column, and diagonal add up to the same number. (Hints: The total of each is 15 and the center number is 5.) One possible solution is given in Figure 4–28.

A magic triangle also uses the numbers 1 to 9. Put them in the triangle so that each side adds up to the same number. One triangle might look like Figure 4–29. Several sums exist that make it a magic triangle.

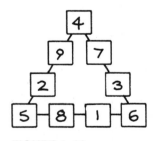

FIGURE 4–29

4. Zooks.   This game can be played with three or four children. Children combine their cards from 0 to 10, shuffle them, and deal them out one at a time until all cards are dealt out.

The purpose of the game is to strengthen the facts for sums of ten. The children place their cards face down in front of them. At a signal, everyone turns over the top card and looks for cards that are turned up that will total ten. The first player to see a total of 10 says "Zooks" and takes these cards. For example, the cards in Figure 4–30 might be turned up.

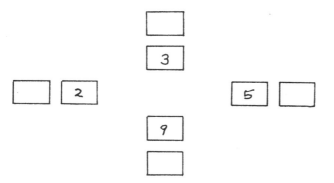

FIGURE 4–30

The player who says "Zooks" claims the 2, 3, and 5. The 9 remains. Players then turn over a new card and again look for sums of ten. If none exist, they turn over a new card. Play continues until one player has all the cards.

   5. *Duel in the Sun.*   This game may be used to strengthen any of the basic facts. It is played by two children at a time. The caller gives each child a number, which is held against the forehead where he cannot see it, but everyone else can. The children stand back-to-back, and take three steps apart. The caller then tells them to turn and gives the operation and the answer. The child can see the other number, but not his own. He figures out his own from knowing the answer and the other number. When the children turn, the caller says, "The product is 40." The first child to guess his number may be the next caller.

Photo 4–6

Photo 4–7

These simple cards may also be used for several other reinforcement games to help children master the basic facts. Children and teachers should make up some of their own. Rules should be kept simple so that memorization of the rules does not interfere with mastering the facts.

Other common games to help children learn basic facts often fall into the following five categories:

1. Dominoes 4. Dice games
2. Bingo 5. Board games
3. Card games

Several variations of these are possible. These five areas were chosen because children are often familiar with the basic rules, and their attention may be focused on the critical aspect of reinforcing basic facts for the four operations. One game for each of these materials is presented here. Teachers and students can think of several more.

| 8-6 | 5 |

FIGURE 4-31a

1. Dominoes. Dominoes are great for any operation. Make a set with the basic facts on one end and an answer on the other. A typical domino may look like the one in Figure 4–31a.

| 18-9 | 9 |

FIGURE 4–31b

Place all the dominoes upside down and mix them up. Each child draws seven dominoes and looks at them. The child with the largest double starts. (For instance, a child with a domino as in Figure 4–31b would start.) The next

	6
	11-2
7 - 1	5	7-2	9	18-9	6	13-7	2

FIGURE 4–31c

child then tries to match one end. In each case, a number question should be matched to an answer. After a few plays, the board may look like that in Figure 4–31c, p. 134. If a player cannot play, he must draw from the pile until he finds a matching domino. The first player to play all his dominoes wins.

2. Bingo. This popular game may be played in a variety of forms. Often the caller calls a number question and the players must find the answer on their cards. (It may also be played so that the caller calls the answer, and the player covers the number question. This version is more difficult, however.) Cards may be made in any size, but 5 × 5 with a "free" space in the center is the most common. Students try to get 5 in a row either horizontally, vertically, or diagonally.

FIGURE 4–32

3. Card games. Card games may be devised on almost any level. A very popular game for children in the early grades is a version of War. Cards are made up with basic facts. The game is often played with two players, although it is possible to play with more. The cards are shuffled, and all are dealt out face down. The players turn up the top card and must work the problem to see who has the higher answer. That player wins the turn and takes both cards. If there is a tie, War is declared. Players then put a card face down and another face up such as in Figure 4–32.

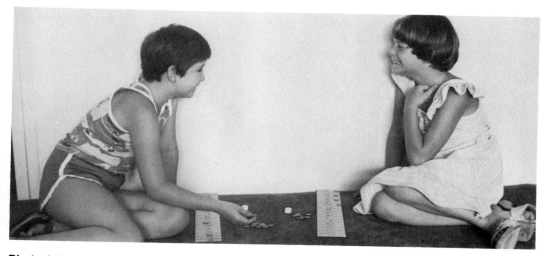

Photo 4–8

The players turn over the center card and the one with the larger sum wins all six cards. Play continues until one player wins all the cards.

4. *Dice games.*   Simple dice games may be devised to reinforce any of the basic operations. Cover-up is a good game for addition of three one-digit addends. Three dice and a board numbered 1 to 18 are needed for each player. Players take turns rolling the dice. A player may cover either the sum of the numbers on the dice rolled or any addends that add up to the same sum. The first player to cover all 18 numerals is the winner.

5. *Board games.*   Many commercial board games exist that help children with basic facts. The teacher or the children may wish to make their own version of one of these. A simple one to make, Islands, requires a board with islands of numbers and two dice. It can be used for any operation, although the example here shows one variation for multiplication.

Each child starts with his marker on "Start." Children take turns rolling the dice. If the product of the two dice is on the first island, the player moves to that space. If not, he does not move in this turn. Children progress from island to island in order. All players try to reach the last island.

Keep in mind that the games mentioned here are designed to reinforce basic facts *after* the children understand the concept behind them. It is important that facts not be memorized by rote with no understanding.

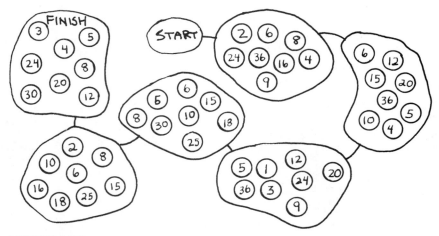

FIGURE 4–33

Extending Children's Skill

As children progress beyond the basic facts, the place-value concepts developed in chapter 3 become increasingly important. Many of the problems children have with addition and subtraction algorithms can be traced to lack of understanding of place value.

*Children Adding and Subtracting with and
Without Regrouping*

The same materials used for the developing of place-value concepts may be used for the concrete development of addition and subtraction algorithms. These may include Multibase Arithmetic Blocks, bean sticks, buttons, chip trading, and the abacus. At this point, most work will be done with the decimal numeration system. This corresponds to the addition and subtraction algorithms being developed.

As children develop the addition algorithm for problems with addends greater than ten, they build on the following:

1. The meaning of addition
2. The concrete understanding of place value
3. Methods of recording addition problems

With this background, addition with regrouping becomes a natural extension of earlier concepts.

The following activities help this development. With each material, children should keep a written record of their work.

1. Base ten blocks. Once children are able to represent numbers such as 28, 232, and 4,896 with the base ten blocks, they are ready for addition number questions. Give children such questions as 25 + 62 and 83 + 14, and let them use the blocks to find the sum. Try 25 + 38. If children produce the results as in Figure 4–34, the array of blocks should be studied to suggest the possible exchange on the 13 unit blocks.

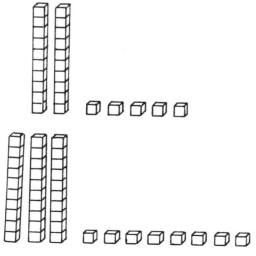

FIGURE 4–34

Photo 4–9

If children are unable to figure it out on their own, they may be asked if there is another way to represent 13 ones. Can a trade be made? The children should be able to trade ten units for one ten for an answer of 6 tens and 3 ones or 63. Let children work several number questions of this type and record their results. Children may produce recording systems of their own. If not, the teacher might want to suggest a simple one, such as:

$$
\begin{array}{r}
25 \\
+38 \\
\hline
13 \\
50 \\
\hline
63
\end{array}
$$

      13   (initial sum of the ones)
      50   (initial sum of the tens)
      63   (sum after trading)

Children should be able to work addition number questions with three or four-digit addends, once the regrouping process is understood. They often enjoy the challenge of working with larger numbers. Encourage children to bring in "real-life" problems that require them to add large numbers. Application enhances reinforcement.

2. Bean sticks. Solutions using bean sticks require many single beans for units, tongue depressors with ten beans glued on for tens, and sets of ten tongue depressors with beans bonded together for hundreds. Use the same type of number questions as those with the base ten blocks. Again, encourage children to keep a written record of their work.

3. Buttons, tongue depressors, beads, etc. Any material that can be grouped by ten may be used to help children with addition. Ten buttons may be put in a small plastic bag to represent ten, and ten plastic bags combined for a hundred. Ten tongue depressors may be joined by a rubber band to represent ten, with ten groups of ten for a hundred. A string with ten beads may represent ten, with ten strings for one hundred. These materials may be used like the Multibase Arithmetic Blocks or bean sticks to solve number questions.

Children may want to try collecting things. Trying to collect 10,000 bottle caps reinforces addition and place value, as children devise methods of record keeping. Can they make a bar graph? How many hundreds will it take? What is the total each day? Can everyone group tens, hundreds, and thousands? How should it be done?

4. Chip trading. The materials mentioned previously have been proportional place-value materials. A "ten" is ten times as large as a "one," and a "hundred" is ten times as large as a "ten." Chip trading and the abacus are nonproportional. The size does not give a clue to place value. In chip trading, the color of the chip indicates its value. On an abacus, it is the position or the rod that shows the value.

Commercial chip trading activities suggest several ideas for developing place-value and operational concepts. Any material with differing colors will work for these activities. Colored 2 cm cubes are a good substitute.

Let children make up a scheme such as:

$$1 \text{ red } = 10 \text{ greens}$$
$$1 \text{ blue } = 10 \text{ reds}$$
$$1 \text{ yellow } = 10 \text{ blues}$$

Children may work problems on a place value chart such as:

	Yellow	*Blue*	*Red*	*Green*
		2	3	4
+		8	6	7
Before Trades		10	9	11
After Trades	1	1	0	1

Children may also enjoy banking games or dice games that have dice color-coded to match the cubes. Using the color scheme above with a red and a green die, children attempt to get a yellow by rolling the red and green dice. If a child rolls 6 red, 6 green and then 5 red, 6 green, trades are made, and the child ends up with 1 blue, 2 red, 2 green. Children take turns rolling the dice, and everyone tries for a yellow by eventually collecting enough cubes to trade ten blue for one yellow. If the child keeps track of each play, the foundation for the addition algorithm is strengthened.

5. *Abacus.*   There are many types of abacuses on the market. One of the most useful for help in developing addition and subtraction algorithms is the ten-bead, closed abacus, which is similar to the one in Figure 4–35.

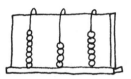

FIGURE 4–35

The tenth bead is the same color as the beads on the next wire. This reminds the child that when he gets ten beads on one wire he may trade them for one bead on the next wire. Children should have a great deal of practice in representing numbers on the abacus before attempting to use it for addition. Some children will find it difficult to keep track of the second addend, as they work problems. Efficient record keeping may help them. The abacus is generally a more complex aid than the materials mentioned previously. It is suggested that the other materials be used to develop the concept before the abacus is used.

Many children at ages seven to nine will have the Piagetian concept of reversibility. They will be able to see subtraction as the reverse operation of addition. They can undo the addition process for subtraction. They understand that if 10 ones = 1 ten then 1 ten = 10 ones. This understanding is important to the development of the subtraction algorithm, especially if regrouping is involved. If children have these concepts, then the same materials used to develop addition algorithms may be used in the development of subtraction algorithms.

When children are encouraged to use materials, they may or may not discover the traditional decomposition subtraction algorithm. This is the algorithm that uses regrouping of the minuend to facilitate the subtraction of the subtrahend. For example,

$$\begin{array}{r} 235 \\ -178 \\ \hline \end{array} \qquad \begin{array}{r} 2\overset{1}{\cancel{3}}\overset{2}{\cancel{3}}\overset{1}{5} \\ -1\,7\,8 \\ \hline 5\,7 \end{array}$$

If the children discover other workable algorithms through the manipulation of materials, let them use those. They may learn the more traditional method later. The following are some methods children may use in solving subtraction problems that require regrouping. Children should record their work with each material.

1. Base ten blocks.   Give the children a problem such as: "John has 53¢. If he buys apples costing 38¢, how much change should he get?" Let the children find the solution using the blocks. Different children may use different solutions. Let them discuss their methods.

One method may be using the multibase blocks, as in Figure 4–36.

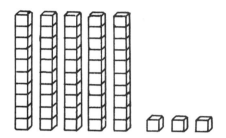

FIGURE 4–36.   Take 5 longs and 3 units.

From this are needed 8 units, but there are only 3. Trade a long in for 10 units. There are now 4 longs and 13 units.

This can be written as shown in Figure 4–37, 5 tens 3 ones = 4 tens 13 ones.

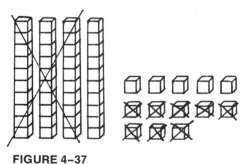

FIGURE 4–37

From this can be removed 8 units and 3 tens. There are 1 ten and 5 ones left. Therefore the change should be 15¢.

This can be shown as:

$$
\begin{array}{c c c}
5 \text{ tens } 3 \text{ ones} & & 4 \text{ tens } 13 \text{ ones} \\
-3 \text{ tens } 8 \text{ ones} & = & -3 \text{ tens } 8 \text{ ones} \\
\hline
& & 1 \text{ ten } 5 \text{ ones}
\end{array}
$$

2. Bean sticks. Other children may do the same problem in a different way using the bean sticks.

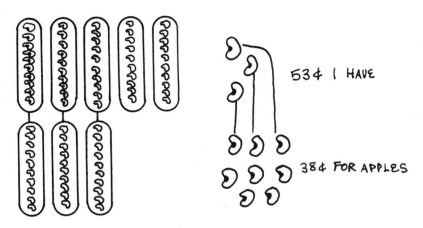

FIGURE 4–38

Compare the two. Three tens and 3 units can be matched. There are 2 tens left from which it is necessary to make a trade so that there are 1 ten and ten ones remaining. Comparing this to the remaining 5 ones, there are 1 ten and 5 ones left. There should be 15¢ change (Figure 4–39).

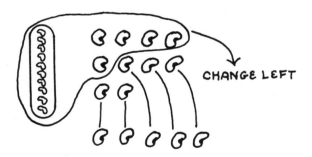

FIGURE 4–39

3. Chip trading. Other children may do the problem in yet another way using chip trading. Let blue = 10 red, red = 1. If the apples cost 38¢, that is 3 blue and 8 red. Fifty-three cents is 5 blue and 3 red. This can be written:

38¢	=	3 tens	8 ones
+ ☐		+ ☐ tens	☐ ones
53¢		5 tens	3 ones

Photo 4–10

Fewer than 2 more blues are needed, because 2 blues would result in 5 blue and 8 red, which is too much. One blue will be selected. Now there are 4 blue and 8 red (Figure 4–40).

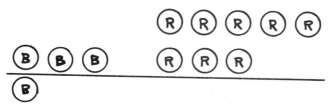

FIGURE 4–40

There are still not enough. If 2 more red are taken, there will be 10 red, which may be traded for another blue. This results in 4 blue and 10 red, or 5 blue. To get 53¢, 3 more red are needed. It is necessary to take a total of 1 blue and 5 red, or 15 more. Therefore, the change will be 15¢.

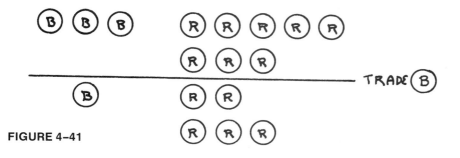

FIGURE 4–41

These three are only examples of the many ways children may solve this problem. All are mathematically sound, and all can be recorded as the children work. Encourage this creative process. Later children may learn more traditional algorithms for purposes of speed, if they wish, but early rote learning of algorithms can destroy the important process of understanding.

Using Diagnostic Teaching Techniques

Because operations with whole numbers occupy such a major portion of the mathematics program during the first three years of school, children having problems in this area are a major concern. Too often, children who make mistakes are told to redo a problem or to work more problems of the same type, with no offer made to determine the source of the errors. Reworking incorrect problems may reinforce incorrect methods. The teacher should discover the cause of the problem. The process must be corrected, not just the result.

There may be several reasons that a child is making mistakes. These include:

1. Social, physical or emotional problems
2. Lack of prerequisite skills or stage of development
3. Weak knowledge of basic facts
4. Incorrect or incomplete algorithm
5. Wrong operation

Examples of these are:

1. Social, physical or emotional problems. Children may be hampered in cognitive skills by noncognitive problems, such as short attention span, hunger, fear of reprisal for getting the incorrect answer or not completing work, or the desire to be "like everyone else."

2. Lack of prerequisite skills or appropriate stage of development. Children may not be ready to learn a particular concept, because they have not mastered previous concepts, or because they have not reached the appropriate developmental stage. A child may not be able to master long division, because he does not understand the multiplication, which is a subskill of division. Another child may not be ready to learn missing addends, as the inverse of addition, because the child has not reached the concrete stage of development and does not have the reversibility concept.

3. Weak knowledge of basic facts. This is probably the most common diagnosis of problems. It may be that children understand the process involved and need further work on more rapid recall of facts. Often children have insufficient understanding of the process involved. Check to see if the children need to return to concrete materials to build the necessary foundations.

4. Incorrect or incomplete algorithm. This is an area in which teacher diagnosis is particularly important. To help in diagnosing incorrect algorithms, ask children to show their work on all problems. If it is not apparent how the children reach an answer, ask them to explain it. Remember that an algorithm is not incorrect just because it is not the one the teacher was taught. Many times algorithms that children have created through manipulating materials are correct, whereas one memorized by rote is not. Try to pinpoint the source of error. Often the mistakes can be corrected by returning to concrete materials and asking the children to record their work as the materials are manipulated.

5. Wrong operation. Children may use the wrong operation by misreading the operation sign or by using the wrong operation to solve a word problem. The latter is a more serious problem. It may be necessary periodically to discuss word problems with children, where they simply explain the operation to perform but do not work the problem. Using real-life situations and problems written by the children helps them understand the appropriate operations to use.

Once children's strengths and weaknesses have been diagnosed, the teacher may wish to group the children for at least part of their instructional time based on this diagnosis. At other times, children with strong skills in a particular area may help children with weaker skills; the class may be taught as a whole group; or the teacher may work with individuals. When grouping, keep the following points in mind.

1. Keep the groups flexible. Do not group in September and expect to have the same groups in June. Groups should change as skills and concepts change.

2. Avoid labelling children. Even if they are called Bluebirds and Robins, children know if the teacher thinks of them as the slow and fast.

3. Avoid giving one group busy work while working with another group. Groups may work independently with materials, games, the textbook, or worksheets, but make sure that the tasks are meaningful.

4. Have interesting tasks appropriate to the level of the group. Each group may have different material, but all materials should be carefully thought out. What may be uninteresting for one group, may be just what another group needs.

Children can learn much from each other. Let them work together in the solution of problems. Children can sometimes explain a solution better than the teacher. The teacher should check on them periodically to make sure the explanations are correct.

Children's Use of Hand-held Calculators

With the availability of simple minicalculators selling for $5 to $10, more and more families are purchasing them. As children use calculators at home, it becomes necessary to examine their use in the classroom.

Calculators can be a useful tool in many areas of an early childhood mathematics curriculum. They do not negate the need for the child to learn basic computational skills. Calculators should help strengthen these skills. Many commercial calculators exist for the sole purpose of strengthening children's rapid recall of basic facts. These are programmed to present children with basic facts and to check their responses. Simple calculators with children asking a number of questions and the calculator giving responses are briefly discussed. These calculators may be used to develop or strengthen skills in areas, such as those listed.

1. Development of problem-solving processes. The calculator de-emphasizes the mechanical aspects of arithmetic, and allows children to concentrate on other aspects, such as analyzing the problem, selecting the correct calculations, and estimating the appropriateness of the answer. Several games have been developed that employ problem solving strategies.

Photo 4–11

2. Recognition of patterns. Working with a calculator often allows patterns to appear more quickly than when working with paper and pencil. Children may develop rules and check them out on a calculator. Children may predict that multiplication is repeated addition and then try it. Other children may develop the commutative or associative property and then test it using several examples. Patterns and rules are an integral part of any mathematics program.

3. Estimation skills. When presented with a number question, children should always estimate the answer before working it on the calculator. This develops estimation skills, as well as checking for errors in the use of the calculator. As children get older, they will often work problems that are beyond the scope of the calculator. For these problems, children must know how to round off and to estimate.

4. Check of calculations. This is a very common use of a calculator. Children may work out solutions using paper and pencil and then check them with the calculator. If errors are found, children then need to rework the solution. In this way, children can work several solutions and get immediate feedback. The teacher is free to work on other aspects of teaching.

Several books with suggestions for developing problem solving skills with the calculator are listed at the end of this chapter.

Extending Yourself

1. Analyze several children's worksheets in addition or subtraction. Identify one child who is having problems and interview the child individually. Pinpoint the problem, and set up a plan for correcting it.

2. Because of a strong "Back to Basics" movement, some teachers feel that mathematics time is best spent in drill on basic facts. Do you agree or disagree with this? Support your position. Discuss your position with someone who has an opposing viewpoint.

3. Choose one operation and a grade level in which it is taught. Look at three or four different textbooks for that grade, and determine how the operation is taught. Which method or book do you prefer? Compare your opinion with a classmate's.

4. Choose one concrete material, such as Cuisenaire rods or multibase blocks, which you could use to teach the concepts of the basic arithmetic operations (addition, subtraction, multiplication, and division). Develop a series of lessons using that material to teach the concept of one of the operations.

5. Describe at least three algorithms you could use to find the sum of 86 and 95. Describe an algorithm that would not work. How could you help a child who might be using this wrong algorithm?

6. List five concepts that would be useful to a child learning basic multiplication facts. Explain how each concept would help.

7. What is the relationship of union of sets to addition? How are they alike? Where do they differ? Describe a series of lessons that would move from union of sets to addition of whole numbers.

Bibliography

Ashlock, Robert B. *Error Patterns in Computation*. Columbus, Ohio: Charles E. Merrill Publishing Co., 1972.

Brownell, William A. and Moser, Harold E. "Meaningful Versus Mechanical Learning: A Study in Grade III Subtraction," *Duke University Studies in Education*, 8 (1949): pp. 1–207.

Coxford, Arthur. *Uses of the Calculator in School Mathematics, K–12*. Monograph. Lansing, Michigan: Michigan Council of Teachers of Mathematics, 1976.

Davidson, Jessica. *Using the Cuisenaire Rods: A Photo Text Guide for Teachers*. New Rochelle, New York: Cuisenaire Company of America, Inc., 1969.

Gibb, E. Glenadine. "Children's Thinking in the Process of Subtraction," *Journal of Experimental Education*, 25 (September, 1956), pp. 71–80.

Immerzeel, George. *Ideas and Activities for Using Calculators in the Classroom*. Dansville, New York: The Instructor Publications, Inc., 1976.

Jacobs, Russell F. *Problem Solving with the Calculator*. Phoenix, Arizona: Jacobs Publishing Co. Inc. 1977.

Reisman, Fredricka K. *A Guide to the Diagnostic Teaching of Arithmetic*. Columbus, Ohio: Charles E. Merrill Publishing Co., 1978.

Robinson, Ann, Ed. *The Hand-Held Calculator*. Monograph. Iowa: Iowa Council of Teachers of Mathematics, 1976.

Thiagarajan, Sivasailan, and Stolovitch, Harold D. *Games with the Pocket Calculator*. Menlo Park, California: Dymax, 1976.

Van Engen, Henry and Gibb, E. Glenadine. *General Mental Functions Associated with Division*, Educational Services Studies, No. 2. Cedar Falls, Iowa: State Teachers College, 1956.

Zullie, Mathew E. *Fractions with Pattern Blocks*. Palo Alto, California: Creative Publications, Inc., 1975.

C H A P T E R

Children's Space

Children Exploring Space

Just as the typical three-year-old child has considerable prenumber experience because of exploration and play, he also has developed a foundation on which to build spatial concepts. He has explored space by initially thrashing about in a crib or playpen and crawling toward objects or open doors. Children discover that some objects are close, while others are far. They discover that various objects like rooms or building bricks have boundaries, and sometimes, if a door is left open, the boundaries can be crossed. They discover that certain items belong inside boundaries, like father's nose belongs within the boundaries of his face, or the bathtub belongs within the confines of a bathroom.

Children also discover that events occur in sequence or order. Early in their lives, they have learned that their own crying is often followed by a parent appearing and their being attended to. Later, children notice that a stacking toy is put together with certain parts in a particular order.

Each of these examples illustrates children's initial experiences in space. They are far removed from school experiences with geometric shapes but nonetheless illustrate a Piagetian discovery about the nature of early spatial development. Research suggests that children learn first about the common objects in their environment; then about shapes through a topological perspective; and, finally, about shapes typically taught in school, called Euclidean shapes.

Photo 5–1

Discovering the Environment

Between the years three and five, children exhibit considerable ability to learn from and about their environment. They play with blocks or bricks, which they can easily differentiate from one another, and they build rooms in which people are placed, garages in which cars are kept, or beds in which babies are found. They sometimes build "rooms" large enough for themselves to fit. Towers are built. Trains are constructed. Comparisons are made.

The role of parent or teacher is one of facilitator. The following activities show how children can be encouraged to expand their spatial awareness, concepts, and language.

1. Encourage children to build. If children build a house, encourage them to talk about it. Who does it belong to? How many are in the family? Can the children show you where each family member sleeps? Oftentimes children will construct things without needing adult suggestions and will only be burdened by a barrage of questions. Be aware of children's needs and interests in building.

2. Provide many materials that can be used to make constructions. Children ages three to five benefit from such materials as building bricks, wooden shapes, tiles, milk cartons, large logic blocks, nesting cups or cubes, and stacking rings. Accessories such as dolls, model cars, and animals help motivate construction of pens, homes, garages, and other such structures.

3. Use language that denotes aspects of space such as position (*up, under, inside*), volume (*full, huge, some*), order (*comes before, between, after*), and comparison (*larger, shorter, bigger*). Discuss where things belong and how they can be easily put away. Move about and put materials away.

4. Gather two each of several common space figures, such as a ball, a large ring (perhaps from a stacking toy), a cube, and a can. Let children feel an object behind their backs or by reaching for it in a bag or box. Have the children see if they can select the one they are holding from three or four other, duplicate objects. If this activity appears too difficult, try the same activity using items from the children's world, such as a fork or spoon, a pencil, a book, or a toothbrush. Later, return to the space figures.

5. Body movements provide experiences that help children explore and define space. Have the children reach up as high as possible, out as far as possible, back as far as possible. Have the children become as small as possible, then as large as possible. Have the children become as fat as they can, as thin as they can. Have the children echo-clap, with the teacher beginning; then have the children begin. Have the children turn all the way around, part way around. Have the children balance on one foot with their arms out, then again with their arms at their sides; then have them move their arms from their sides until they are stretched out. Encourage the children to climb on classroom apparatus, stairs, climbing bars, and small slides.

6. Play Follow the Leader around the room with various children circling, crawling, and hopping. Use music to accompany the action.

Through their physical activities, children develop an understanding of the dimensions of their space. Sensory perceptions continue to influence how children think about space. But changes begin to occur. These are discussed later as children develop size and shape constancy.

Topological Space

The perceptions of three- to five-year-old children are topological. *Topology* is the study of space concerned with position or location, where shape and length may be altered without affecting a figure's basic properties of being open or closed. For example, if a four-year-old is shown a triangle and asked to make several copies of it, he will likely draw several simple closed curves but not necessarily triangles (see Figure 5–1). To the child, all of the drawings

FIGURE 5–1

are the same, because he perceives that the triangle only has the property of being closed (younger children often draw figures that are not closed). Thus, a triangle may be "stretched" into any closed figure (see Figure 5–2).

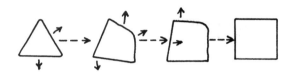

FIGURE 5–2

The study of space in which the figure and any enclosed space must remain rigid or unchanged is called *Euclidean geometry*. The historical development of geometry was Euclidean; that is, it developed from figures such as points, lines, and polygons. Some of the research of Jean Piaget has implied that children do not develop geometric concepts in a Euclidean manner. Because of their topological perspectives, young children need the active, exploratory period suggested by the activities in the previous section.

In chapter 3, relationships with numbers were discussed as the concept of number was developed. Likewise, spatial relationships can be identified as the concepts associated with space are developed. Children who perceive the world from a topological point of view are developing an understanding of four basic relationships:

1. "Is close to" or "is far from"
2. "Is a part of" or "is not a part of"
3. "Comes before" or "comes after"
4. "Is inside of," "is outside of," or "is on"

1. In the world of the three-year-old, "is close to" or "is far from" might determine what the child plays with. An object or toy near at hand is a part of the child's immediate world and is likely to be picked up and played with, whereas an object far away may be temporarily unimportant. At this age, children consistently distinguish between objects near and those far away.

When children draw a picture of a person, they put the fingers close to the hand and far from the shoulder.

2. As they develop, children begin to separate objects visually. Thus, they find their bedroom furniture "is not a part of" a wall. This is confirmed when their room is rearranged. Also, their toys are not a part of the toy box. When children draw a picture of a person, the arm is separated from the lower part of the body; the feet are separated from the head.

3. Understanding of sequence develops as children expect certain events to "come before" or "come after" other events. The infant soon learns that a shrill cry is quickly followed by sounds of an approaching parent, the appearance of that parent, and attention by the parent. Three- to five-year-olds learn that waking up comes before getting dressed, which comes before eating breakfast. They learn that the engine comes before the passenger car that comes before the caboose on their toy trains. In their drawings, children often will draw a person by drawing the head first, followed by eyes, nose, mouth and then perhaps hair or ears and the rest of the body.

4. As children discover that a particular toy or object "is inside of" or "is outside of" a box or an area of a room, they are developing an understanding of boundaries. Children readily discover through group games about being inside the circle of children or by chasing one another around the outside of a circle. They know that beds belong inside the house and that swing sets belong outside the house but inside the yard. Children also distinguish between standing on a boundary and being within or without the figure formed by the boundary. In their drawings, children make eyes, nose, and mouth within the boundary of the head.

The following activities provide three- to five-year-olds with beginning spatial (topological) experiences.

1. Play "Who Is Closest To?" or "Who Is Farthest From?" Have each child seated in particular locations within the room. Ask, "Who is closest to the door?" (carpet, window, sink, piano, building-blocks, etc.). Have that child stand. In some cases, it may be helpful to have the other children identify who is the closest to a particular object. Change the question to, "Who is farthest from" a particular object. Encourage the children to ask the questions.

2. "Find" is an activity that helps children think about the characteristics of objects. The children are separated into groups of six or seven and sit in a circle within which is a string or yarn loop. Inside the loop are several objects, such as straight and crooked sticks, sharp and dull pencils, large and small wooden cubes, pieces of paper and sandpaper, a marble, and a large ball. Children are asked to pick out certain objects. For example, Sue may be asked to pick out a large, round object. Dave may be asked to find a short, crooked stick. Objects may be compared. A crooked stick might be compared with the straight stick; the small cube with the large cube. Many possibilities exist for strengthening children's perceptions of the characteristics of objects. Later, children will distinguish among the characteristics of geometric shapes.

Photo 5–2

3. On the floor or playground surface, chalk out several large closed curves. Have groups of four to six children follow directions with respect to the figures. Have them stand inside the figure; have them stand on the boundary; have them walk on the boundary forwards and then backwards. Have the children stand outside of the figure; then have some inside; finally, have some outside and at least one child on the boundary. Have two children stand in the figures such as those indicated in the drawings at A and B (Figure 5–3). Ask one of the two children to walk to the other, following one rule: they cannot cross over a boundary line.

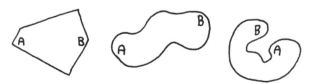

FIGURE 5–3

4. Provide simple picture puzzles for the children to complete. Some puzzles might have the outline of each piece drawn on a piece of cardboard or in the puzzle frame, so that the less able children could complete the puzzle. Other puzzles would not have the outlines available.

5.   Have the children play circle-chase games. The whole group forms a circle, standing and facing the center. One child who is "it" begins walking around the outside of the circle formed by the other children. He then taps one of the circle players on the back and races around the circle, while the tapped child chases and tries to tag him, before he can get to the spot vacated by the child tapped. If the child who was in the circle and was tapped tags the child who was "it," the circle child maintains his place in the circle; and the child who was "it" must tap another child. Variations of this game instead of tapping would be for a handkerchief or piece of cloth to be dropped or calling the name of the circle child.

6.   Play a game called "Fits Into." See if children can determine if various objects fit into certain containers. Would an elephant fit into a shoebox? Would a pencil fit into a paper bag? Would a can fit into a drinking glass? Would a child fit into a house? Would a tricycle fit into a lunch box? Later, see if children can think of objects that would or would not fit into containers.

7.   Allow the children to draw and paint. Provide the materials and space to encourage creative work. Note the relationships in the drawings, how the body parts are joined, and their relative positions.

Photo 5–3

Children's Art

A clearly distinguishable extension of children's topological thinking is found in their art work. Brief examples of children's drawings were cited earlier in the discussion of the topological relationships that are readily perceived by younger children. Photo 5–3 shows drawings by young children that illustrate those topological relationships with respect to perceptions of the human body.

Children's Sand and Water Play

Children's initial understanding of volume and displacement develops from their play. Very early, children play as they bathe. They enjoy water toys including small animals, boats, and assorted containers with which they can dip and pour water. It is through such informal play that youngsters develop an intuitive concept of volume.

　　Water and sand provide experiences in filling and emptying containers of various sizes, comparing containers for equal amounts, and estimating. Clay, papier-mâché, and play dough similarly offer experiences in constructing and comparing amounts. The nature of those latter materials are, however, dif-

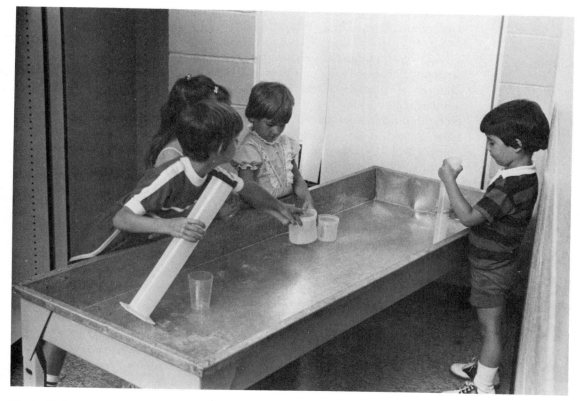

Photo 5–4

ferent from sand and water in their manipulative characteristics. The following are some typical activities that use sand and water for developing a foundation for understanding volume or capacity.

1. Make available to youngsters a low sink or water-tight box. Provide containers of various sizes and shapes, such as cans, milk or juice containers, plastic cups, or glasses. Allow the children time for free play. The adult's role at this time is to let the children play undisturbed. Interaction might take place, if the children were to call the adult over to the water box to make a comment or ask a question. The same type of activity should be available at a sand box.

2. Provide the children with one particular container to use as the pouring container and several others to have water (or sand) poured into. Ask the children to identify the containers that can be completely filled when they start with the pouring container completely filled. Put those that can be filled together and those that cannot be filled together.

3. Have the children take a small container and use it to fill a larger container. Later, ask them to guess how many times they think they will need to pour from the smaller into the larger container in order to fill it. Have them count the number of times it takes.

The years three to five are important in children's development of spatial concepts. The activity is rarely formal. It is an extension of their everyday play. Children are encouraged to explore space by moving about, building, and comparing. The media are varied. Children expand their experiences with everything from blocks to water. As the understanding of spatial relationships expands, so does the language used to communicate those relationships.

Euclidean Discoveries

Between the ages five and seven, most children develop the ability to understand the meaning of Euclidean space. That is, the children develop their ability to reproduce shapes without significantly altering the characteristics of those shapes. For example, in the previous topological stage, children would copy a figure but would allow corners to become round and distances to change. At the stage of Euclidean understanding, corners remain corners and distances are unchanged—the figure is considered rigid.

The shift from topological to Euclidean thinking is not sudden. It may occur over a period of two years. Thus, usually between four and six, children can recognize and name the more common figures: line, square, triangle, rectangle, circle, star. Other figures are neither identified nor differentiated from these shapes. For example, the rhombus and square may be confused, as might the rectangle and parallelogram. Even more difficult for children is copying various shapes from blocks or drawings. Children may be able to identify accurately shapes long before they are able to produce their own examples.

Continued Development in Topological Concepts

It is necessary to continue activities that relate to topological space for children five to seven. The following are typical of activities that extend topological ideas.

1. Language development in concert with activities is a natural part of teaching. To develop and reinforce practical ideas, such as inside, outside, near, far, on, in, under, over, before, and after, have children sit in small groups at a table on which are numerous objects. The leader gives directions to various children. For example, "Julia, please put the red block as far away from the plastic cup as you can," or "David, please put the short pencil in the tin can." Several children may be participating simultaneously, and the leader can be checking the understanding of the language and the concept. Discussion should be encouraged.

As the children develop skill in reading, the directions for this activity can be written on activity cards, with children correcting one another if errors are made.

2. Games such as the following game that employs boundaries or bases help children understand the idea of inside, outside, between, region, across, and boundary. This game may be called Between the Bases. Three regions are drawn on the playground or floor of a multi-purpose room.

Two groups of children are selected: those who attempt to change from the red base to the green base, when a whistle is blown, and those who begin at the "catchers region." As the children are changing from red to green base, the catchers run from their region and tag those who are changing, as long as they are outside either the red or green bases. The game is over whenever there are no more children to run between the red and green bases.

3. Continue to encourage children to construct and manipulate space figures. Materials might include tiles, logic blocks, geoblocks, cubes, cans, empty milk cartons, Unifix cubes, Cuisenaire rods, Pattern Blocks, Parquetry Blocks, and clay. During play, the teacher may question the children about pictures, buildings constructed, patterns, and shapes. The children can learn to be analytical, when questions are carefully phrased. For example, "Can you make another house just like the one you have made there? I would like you to try." At the same time, the questions can gather information for the teacher.

Photo 5-5

4. Sometimes after children have made constructions, it is useful to have them tell stories about what they have done. When these stories are relayed, the teacher writes what the children say. The writing may take several forms, including: (1) writing on a large sheet of "story paper" so the children may read their stories to one another, while all can see the story; and (2) writing on a sheet of paper or in the children's writing books, skipping lines, so the children might copy the story using the teacher's writing as a model. The language-experience approach helps develop communication skills, as well as developing the children's thinking about space.

5. Provide experiences with boundaries. Earlier, introductory activities were suggested with the children determining whether two children were in the same region. These activities are extended by employing more complex figures. The activity may be constructed on the playground or on paper. When children are physically involved on the playground, they tend to perform better because they become an actual part of the activity.

In Figure 5–4, decide how many regions are determined in each drawing. On the playground the children may want to stand "inside" and see if they can get to the outside by walking.

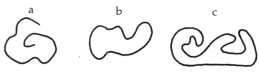

FIGURE 5–4

There is one rule: you cannot step over a boundary line. Thus, in Figure 5–4-a, only one region exists, whereas in parts b and c two regions exist. In Figure 5–5 there are four, one, and three regions, respectively.

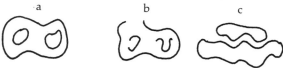

FIGURE 5–5

If these activities are performed on paper, the children might benefit from coloring each region a different color. The clever teacher can devise many variations of this type of boundary activity.

Another type of boundary activity is the maze. The object of this activity is to see if two children are in the same region. On the playground, the children can attempt to walk to one another without walking or reaching across a boundary. On paper, children trace the regions with their fingers. The variations and complexity of designs are nearly unlimited.

FIGURE 5–6

A third, more complex boundary activity involves having the children construct maze puzzles for themselves and other children. Maze puzzles may be constructed by beginning with a simple frame with a "door" to go in and a door to exit, as in Figure 5–7.

FIGURE 5–7

To complete the maze, lines are drawn from any wall. The only rule is: no line can connect another wall. Steps a, b, and, c in Figure 5–8 show how a maze puzzle was constructed. Children find fascination in constructing mazes.

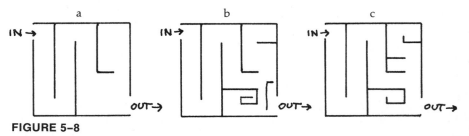

FIGURE 5–8

6. Children should be encouraged to construct picture puzzles. By the age of six or seven, children will not need to have outlines of the shapes to assist them in completing puzzles. Challenge the students with difficult puzzles, and discuss informally with an individual or small group how they have gone about putting the puzzle together. It should be evident that strategies are developed as puzzles are completed. Edges are generally connected first, followed by pieces that form distinct images or have easily matching colors. Pieces are joined when shapes are discovered that fit a region that has been surrounded by other pieces. Finally, all other pieces are put into place by the process of elimination.

Euclidean Shapes

The children's abilities to learn the names and properties of the common Euclidean shapes (triangle, square, rectangle, circle, parallelogram, rhombus, and so forth) will vary considerably within any group of children. Children who are able to observe the shape of an object and easily find another like it, or those who are able to look at a figure and then draw it maintaining the characteristics essential to the figure are ready to proceed with more systematic instruction of Euclidean shapes. Because the children will vary in their readiness, it is important for small group and individual instruction to take place. This transition from topological to Euclidean perspectives should take place during the age range of five to seven years.

Piaget (1953, p. 43) indicated that the learning of shapes requires two coordinated actions. The first is the physical handling of the shape—being able to run fingers along the boundaries of the shape. The second is the visual perception of the shape itself. It is insufficient for children merely to see drawings or photographs of the shapes. There is a wide variety of materials and activities that can help to present Euclidean shapes to children. Below are listed some of these materials and activities.

1. Children are given two-dimensional (flat) shapes with which to play. The shapes may be commercially produced, such as the set of logical blocks, or they may be teacher-constructed from colorful posterboard. The children should be allowed free play with little or no teacher direction. Perhaps they will construct houses, people, cars, animals, patterns, or larger shapes. After having plenty of free time with the shapes, the children will be ready for the teacher to ask a few questions or to compliment them on their work. If a house has been constructed, ask the children to construct some others just the same.

Photo 5-6

Challenge the children to make the same object, except that it be constructed upside down. If a pattern is made, perhaps it can be extended.

2. Construct models of various shapes for the children to handle. One way this may be done is to bend heavy wire in the shape of a triangle, square, rectangle, circle, parallelogram, rhombus. A touch of solder should hold the ends together. Another way is to glue small doweling to a piece of poster board. The children can then develop a tactile understanding of the shapes. Once the shapes have been manipulated, the children can be asked to draw a particular shape while looking at and feeling the models. Later, they may be asked to draw the shapes with only feeling or seeing them. Finally, the children are asked to draw the shapes without either seeing or feeling them.

3. Parquetry Blocks offer a unique way to learn Euclidean shapes. (They are available at toy stores and department stores.) Parquetry Blocks are geometric shapes of varying colors and sizes. The first attempt to use them should be in a free-play activity. Then, there are several ways the blocks may be used to present Euclidean shapes.

Photo 5–7

a. Copy activities include holding up one of the shapes and having children find another the same shape or a different shape. Next, the teacher puts three or four of the blocks together in a simple design, and the children are asked to copy the design. It may be copied exactly or with a slight variation, such as with different colors. Finally, the blocks are again put into a simple design, but the blocks are separated from each other. Copying this design requires the children to visualize "across" the separations.

b. Outlines of Parquetry Blocks are presented, and the children are asked to find the same piece and place it on the outline. Initially, color, shape, and size must be matched. Later, just shape and size are matched. More complicated outlines, using designs of two or more blocks, are presented after the first experiences with single blocks.

c. Children are asked to make their own outlines for others to fill in by either drawing around the various shapes or by putting all the shapes down and drawing around the entire design. This latter variation produces a challenging puzzle for children to complete.

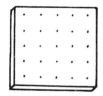

FIGURE 5-9

4. Another dynamic learning aid is the geoboard, which was discussed earlier as an aid in developing early fractional concepts. The geoboard for children ages five to seven would most likely be either a commercially or teacher-make board with a large enough nail pattern for the children to easily manipulate. A board 25 to 30 cm on a side with five rows of five nails each is quite adequate (see Figure 5–9).

After a period of free play to allow the children to discover patterns, shapes, and "pictures" that can be constructed, the first directed activities should be copying activities. The teacher constructs a particular configuration or shape and shows it the to children, asking them to copy it. Initially, a line segment might be constructed, then perhaps combinations of two, three, or

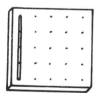 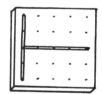 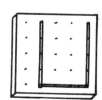

FIGURE 5-10

more line segments (see Figure 5–10.) Next, simple shapes might be constructed, gradually becoming more complex and challenging (see Figure 5–11, p. 166). As soon as the children understand the nature of the copying exercises, allow them to construct shapes for others to copy.

FIGURE 5–11

As the children gain experience in recognizing and naming shapes, the names should be used to describe shapes for the children to construct. For example, "Let's make triangles on our geoboards. If we look at everyone's triangles, can we find some things that are the same? Are there any that are completely different? Who has the biggest triangle? Who has the smallest? Who has the triangle with the most nails inside the rubber band? Who has the triangle with the fewest nails? Now let's make some squares."

The teacher may explore with the children the number of units that make up the boundary of a figure. First, the unit is designated. For example, the distance between two nails in any direction except diagonal is counted as one unit. Second, a figure is constructed and the students are challenged to find how many units there are in its boundary. Next, the children are asked to construct a figure with a certain number of units in its boundary. For example, "Construct a square with a boundary of twelve units" (see Figure 5–12).

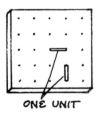 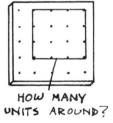 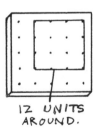

ONE UNIT   HOW MANY
                UNITS AROUND?   12 UNITS
                                     AROUND.

FIGURE 5–12

Simple exercises may be used to have children count the number of squares within a larger square or rectangle. The smallest square is designated as one square unit (see Figure 5–13). Then, the teacher inquires, "How many square units can we find in this figure?" The children would count the number of square units contained in a given figure.

As they gain experience, the children could construct squares or rectangles with a given number of square units, demonstrating initial understanding of a process that eventually leads to full understanding of area or measurement in two dimensions.

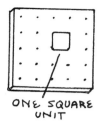

ONE SQUARE
UNIT

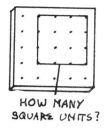

HOW MANY
SQUARE UNITS?

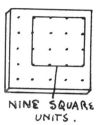

NINE SQUARE
UNITS.

FIGURE 5–13

5. Children gain understanding of Euclidean shapes through arts-and-craft activities. There are several ways to encourage exploration of shapes. The children may be asked to draw a large triangle, square, or rectangle on a sheet of plain paper and design a picture using that shape as a beginning. A variation would be to begin with a shape cut from construction paper and either draw or glue other shapes onto the cutout.

The children may be provided paper with square or triangular patterns. They could then color or outline shapes or pictures as they wish (see Figure 5–14). A variation of this would be to begin with numerous small square, triangular, or circular shapes. The children would glue these shapes onto a sheet of paper to make a picture or design.

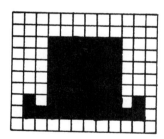
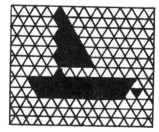

FIGURE 5–14

Tiling a region with a particular shape is called *tessellating.* Of the regular Euclidean shapes, only triangles, squares, and hexagons will completely cover a region without the need for additional pieces to fill in gaps. Early experimentation with tessellations helps children discover characteristics of shapes. There are countless irregular shapes with which a region may be tessellated (see Figure 5–15).

6. As discussed in chapter 2, classification of objects takes place early in the children's school experience. It helps, among other things, to develop a concept of number by extending the notion of objects with the same or differ-

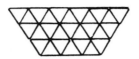

FIGURE 5–15

ent characteristics. Euclidean shapes may be the objects for classification. Thus, "to be the same shape as," "to have the same number of corners as," "to have the same number of sides as," and "to be the same size as" may be used to group various shapes together according to likenesses. Difference and ordering relationships may also be established.

Rediscovering the Environment

Learning from the environment was discussed earlier with children ages three to five. Between ages five and seven children continue to extend their awareness of basic shapes surrounding them. Within the classroom, there are tile patterns on the floor, shapes of books, windows, furniture, light fixtures, and seating arrangements. Outside are shapes of fences, wheels, playing fields, buildings, street patterns, sidewalks, and swing sets. Of course, the number of various shapes children can point out is limited only by their imaginations. Teachers need to encourage children to look around. The following activities should help children gain a fuller understanding of shapes in their world.

1. A class or group of children takes a walk around the playground or neighborhood, searching for a particular shape (for example, squares). Whenever a square is located, it is sketched or recorded for later discussion. Various small groups may each be looking for a specific shape, or the entire class may be looking for all of the shapes they can find. When the walk has ended, the children may make drawings in which the shape is drawn in its original location. The teacher may hold a short discussion reviewing the shapes found and where they were located. The children may write short stories or create a class book using shapes as a theme.

2. The children may be shown pictures of automobiles, homes, or animals and asked to pick out the shapes they find. The youngsters may design their own figures that contain certain shapes. They may wish to see if their classmates can locate the shapes.

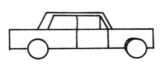

FIGURE 5–16

3. The use of language experience stories and books provides children with the chance to think creatively about shapes. Topics like, "Some Round Things," "Objects That Are Square," or "Finding Triangles" give children the opportunity to consider where in their experiences certain Euclidean shapes may appear. Handsomely illustrated books may result when the teacher allows the children to discover the shapes in the environment.

4. Constructing homes, stores, garages, and the like from cardboard bricks, shoeboxes, or milk cartons helps to remind youngsters of the shapes in their surroundings. The faces of bricks or shoeboxes bring forth the notion that the "faces" of space figures tend to be squares, rectangles, or other polygons. Elaborate playhouses in which the children spend some of their work or play time may be constructed.

5. Shapes in the environment do not necessarily have to be Euclidean in order to be of interest. Shapes of leaves, animal tracks, seed pods, flowers and petals, snowflakes, sea shells, tire tread patterns, landscaping designs, and the like may be used in exploring shapes. Drawings and stories may be used to describe the numerous characteristics observed.

Using Geometric Ideas

Children's growth in understanding geometric ideas is nurtured as they enter the age period seven to nine years. By ages seven to nine, children have moved from the period of topological perception to that of Euclidean perception. That is, shapes remain unchanged, and squares and circles are easily differentiated. Hopefully, children can identify most common shapes. The children should continue their growth through activities designed to build on previous work. Thus, tasks that involve geometry in two and three dimensions and environmental geometry should be pursued.

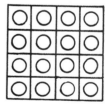

Developing Two-Dimensional Concepts

The first group of activities focuses upon two-dimensional shapes. Later, activities involving three-dimensional shapes (space figures) will be presented.

1. Coloring a quilt pattern presents a challenging puzzle. First, the children are presented an actual quilt or a photograph of one and asked to figure out how the quilt could be colored, so that any two adjoining pieces are a different color. The children might sketch the pattern of the quilt and then color it (see Figure 5–17).

FIGURE 5–17

The second step is to have the children sketch the outline of the quilt and color it using the fewest number of colors possible. The children are to remember they cannot color two adjacent shapes the same color. Four colors will probably be the typical solution for most complex quilts. Some, however, may be colored with two colors and some with three. The children should be encouraged to sketch these quilts. No quilt will ever require more than four colors.

2. Constructing patterns by gluing shapes onto a sheet of art paper gives the children an opportunity to develop interesting and attractive designs as well as to discover how various shapes fit together. For example, if the shapes in Figure 5–18 are cut out, each in its own color, a variety of patterns can be constructed.

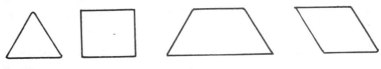

FIGURE 5–18

As mentioned earlier in this chapter, the art of tiling is called tessellating. Tessellation provides children the opportunity to grow further and to experiment with shapes. Any convex quadrilateral (four-sided figure) can be used to tessellate a region. Children should explore this idea by tessellating a small region, such as a sheet of paper with figures like those in Figure 5–19.

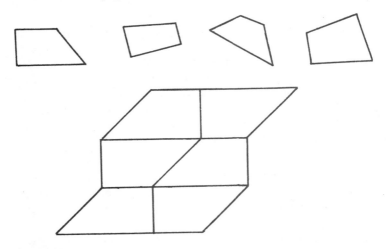

FIGURE 5–19

If twenty-five of a given shape are carefully cut out and glued to paper, the students will be able to produce the tessellation pattern. Children also should be encouraged to find tessellations made of combinations of regular polygons.

3. The simple geoboard can be introduced to five- to seven-year-old children. As they grow in their ability to manipulate the shapes on a geoboard, the children will be able to handle more difficult patterns and activities. The arrangement of nails or pegs provides for the construction of numerous basic shapes. There is the common "square" geoboard on which triangles, squares, rectangles, rhomboids, parallelograms, and other irregular polygons may be constructed.

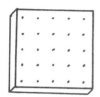 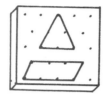 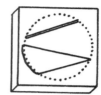

FIGURE 5–20

The triangular geoboard allows children to construct equilateral triangles, as well as a variety of polygons. The square geoboard does not allow equilateral triangles to be made. The circular geoboard is useful for constructing shapes resembling circles, radii, diameters, and figures inscribed in a circle.

Activities could include those mentioned earlier, that is, constructing designs, patterns, and basic geometric shapes. These activities should be expanded by including introductory, measuring activities, so that the children will be determining the number of units in the boundary of a shape or the number of square units within the boundary. For example, "Make as many figures as possible with boundaries that are eight units long. How many square units does each of these figures contain? Make as many figures as you can of four square units. How many units is the boundary of each? Make a square and a triangle with the same number of square units. How can a person find the square units of a nonsquare figure like a triangle?" Which has the greatest boundary, or are they the same?

Unusual shapes may be constructed and the boundaries and areas determined by partitioning off units and square units and counting. Children can create such shapes and challenge their peers (see Figure 5–21).

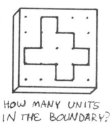  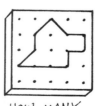

HOW MANY UNITS IN THE BOUNDARY?

COPY THIS FIGURE ON DOT PAPER. MAKE IT TWICE AS BIG.

HOW MANY SQUARE UNITS IN THIS SHAPE?

FIGURE 5–21

With some imagination, teachers may find other ways to use the geoboard. For example, if the distance between any two nails (except diagonally) is designated as two units, three units, four units, or whatever number the teacher wishes, the problem of determining the lengths of boundaries becomes an arithmetic problem. Children may count by two's, three's, or four's, or apply simple multiplication. In addition, determining the area of a given figure offers a challenge. "If each square is equivalent to four square units, can you construct a square of 64 square units on the geoboard?"

4.   The use of tangram pieces was mentioned in the discussion of fractions. Tangrams offer children the chance to engage in puzzles and creative endeavor. There are seven tangram pieces. They may be fitted together to make a square (see Figure 5–22), but such a problem is difficult for many adults as well as for seven- to nine-year-old children.

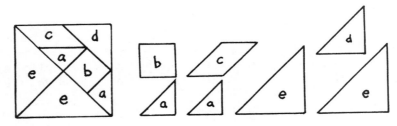

FIGURE 5–22

FIGURE 5–23a

Initial activities should include providing "frames" in which the children fit two or more of the tangram shapes. For example, using an "a" piece and a "d" piece, make the shape as in Figure 5–23a. Thus, the children are able to put the pieces together and achieve success.

As the children gain in their abilities to complete the puzzles, a greater number of pieces is used, and the frames become more difficult to complete. Thus, experienced children might be asked to fill the frame using all but one "e" piece, as in Figure 5–23b.

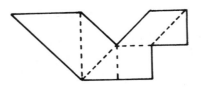

FIGURE 5–23b

Another enjoyable tangram activity is to construct pictures of animals, people, objects, and houses using all or some of the tangram shapes. Children may fill in frames or may construct their own pictures. The "waving man" in Figure 5–24 is an example of such a creation.

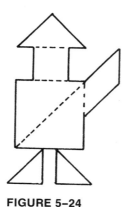

FIGURE 5–24

5. Exploring the Pattern Blocks, Parquetry Blocks, or Cuisenaire rods with mirrors offer children the chance to learn: (1) more about basic shapes and (2) symmetry. Initially, children should be provided with a mirror and several blocks and left to explore. After this period of exploration, a single block is placed on a sheet of paper, and a mirror is put along side or across the top of the block. The children observe the reflections, at times guessing what the image will be. Next, the children place a block on a sheet of paper, on which a line has been drawn to represent where the mirror will be held, and either place another block or draw what they think the reflected image will be.

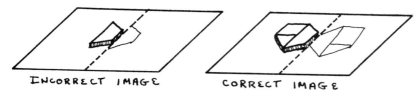

FIGURE 5–25

The mirror is then used to confirm or reject their guess. Soon children are able to construct figures using several blocks and to guess and check their images with the mirror.

Commercially prepared mirror card activities that are designed to lead students through a carefully constructed series of activities relating to symmetry are available. A theme of symmetry may be used to develop a class study of objects and creatures that clearly possess symmetrical properties. Once these have been collected—people, butterflies, buildings, and so forth—displays, bulletin boards, or class books may be constructed.

6. Exploring pentominoes offers children the opportunity to test perceptual and creative abilities. Pentominoes are configurations produced by combining five square shapes of the same size. There is one rule: each square shares at least one complete side with another square in the configuration. Three of twelve possible pentominoes appear in Figure 5–26.

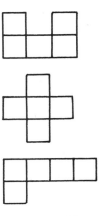

FIGURE 5–26

Figures are two-dimensional and are considered to be the same, if one is a flip or rotation of another. For instance, those in Figure 5–27 are the same piece.

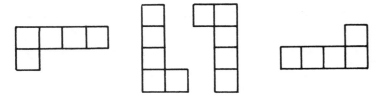

FIGURE 5–27

Initially, children should be given numerous square shapes to explore. They may be challenged to see how many configurations they can produce. For seven- to nine-year-old children, squares of 3 cm on a side are ideal. As the children discover the patterns, they may shade or color the patterns on a sheet of graph paper that is marked off with squares of 3 cm.

Another way to investigate pentominoes is to use the small milk containers so common in schools. The top of each container is cut off so that the bottom and four sides are all the same sized square shapes. Then, the children are asked to see how many of the twelve pentominoes can be made by cutting the cartons along edges and without cutting any one side completely off. Figure

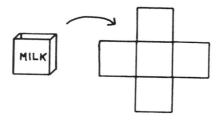

FIGURE 5–28

5–28 shows one example in which cuts were made along each of four vertical edges, and the sides were folded down.

Once children are comfortable with pentominoes, they may wish to try to tessellate with various pentominoes. When each pentomino is used repeatedly, is it possible to always cover a sheet of paper without leaving gaps? (See Figure 5–29.)

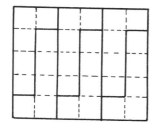
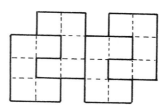

FIGURE 5–29

Pentominoes may also be used to explore symmetry further. Can a mirror be placed on all or some of the pentominoes so the reflection is the same as that part of the figure behind the mirror? In other words, do all pentominoes have line symmetry?

7.   Visual perception is strengthened as children search for shapes in complex figures. For example, in Figure 5–30 children may be asked to find the number of triangular shapes (8) in 5–30a; the number of square shapes (11) in 5–30b; or the number of triangular (6), rhomboid (6), or trapezoid (6) shapes in 5–30c.

a   b   c

FIGURE 5–30

The geoboard provides the opportunity to search for regular and irregular figures. How many squares can be constructed on the geoboard? How many triangles? How many trapezoids? Is it possible to construct a regular pentagon or hexagon (all sides the same length) on a geoboard?

8.   An angle may be thought of as occurring with a change in direction along a straight line. Consider walking along a line and coming to a place at which the line changes direction, however slightly or sharply (see Figure 5–31). The combined path is an angle with the corner being called a vertex.

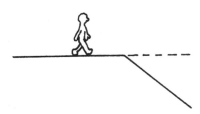

FIGURE 5–31

Children studying the characteristics of plane figures will soon be discussing the corners or bends that help to enclose a figure into whatever shape has been produced. The concept of angle is being developed at an intuitive level. To firm up the concept of an angle, children should be given lines of chalk or tape on which to walk. At first, a line might just change direction every few meters forming a zig-zag path. Later, the figures may be large polygons. The children can get into the habit of using their arms to indicate the angle. This is done by having them point with their arms in the direction they are walking. When they come to a change of direction, one arm points in the direction in which they had been walking; the other arm points in the direction of change. Thus, they are forming the angle of change with their arms (see Figure 5–32).

FIGURE 5–32

Observation and discussion should soon focus on size of angles. It will be noticed that some seem to be "square corners," while others are more or less than "square." The geoboard can be a useful learning aid on which to construct angles of varying sizes. Right, acute, and obtuse triangles may well emerge as an outgrowth of constructing various sized angles.

Increasing Spatial Awareness

Up to this point, the activities have dealt principally with plane figures—figures of two-dimensions. As do all individuals, children live in a three-dimensional world. Their movements, explorations, and constructions have been, since their earliest age, in space. The exploration of space is the classic example of early mental growth. As children continue their growth in geometry, activities with three-dimensional space figures are an important part of this learning. Whenever possible, the children's environments of the classroom, home, and community should be tapped. The activities that follow are designed to add to the growth of spatial concepts. Again, activities cannot by themselves teach. They must be augmented by reading, discussion, example, and thought.

1. The schoolyard or neighborhood walk was mentioned in describing activities for children ages five to seven. It is also effective with older children and can challenge their imaginations. On the "shape walk," the students may be asked to sketch the shapes they observe. The shapes may be two- or three-dimensional. That is, the shape of windows, doors, faces of bricks, or fence patterns may be sketched; or the shape of entire homes, individual bricks, garbage cans, or light posts may be sketched. It is likely some discussion will be needed regarding how to sketch three-dimensional figures. The children should share with one another their own techniques. Descriptive stories by the class or individuals may help to culminate an investigation of shapes in the community.

2. Children continue to benefit from construction with three-dimensional objects. The materials may change from large blocks and building bricks to more compact materials such as tiles, Geoblocks, Unifix Cubes, Cuisenaire rods, and Pattern Blocks. The goal remains the same—to allow the children to explore space. Beyond exploring, the children may construct figures for others to copy or to produce a figure and a mirror image of the figure. Figure 5–33 and its mirror image was constructed using the Cuisenaire rods. A

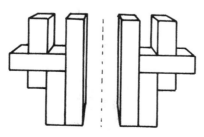

FIGURE 5–33

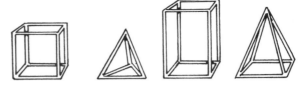

FIGURE 5-34

challenging series of work card activities, accompanies the set of Geoblocks. The teacher may wish to develop other, similar cards for use with the other three-dimensional learning materials.

3.   Straws and pipe cleaners (or straws of two sizes) are used to construct space figures. Initially, two-dimensional figures may be produced; and, as space figures are investigated, it should become apparent that the faces of virtually every space figure except the sphere is a polygon. Thus, when a cube is constructed, an investigation of its faces will yield squares. If a tetrahedron is constructed, an investigation of its faces will yield triangles. Children should be encouraged to construct various polyhedra. Several are shown in Figure 5-34. Once numerous space figures have been produced, the children should

Photo 5-8

compare them, noting the number and shape of faces, number of vertices, and interesting facts about their shapes. These findings may be recorded in written or chart form and prominently displayed.

4. Projects using space figures offer motivation for creative learning experiences. One such project was initiated during an introductory class on space figures. As the children and teacher looked at a set of Geoblocks, one child noted that a particular piece looked like an Egyptian Pyramid; another student thought that the word *prism* sounded like *prison.* The following day, the teacher arrived at school with several slides from the 1967 World Exposition at Montreal, "Expo—Man and His World." A boy in the class mentioned that it would be exciting to create a city full of shapes. The Geoworld project was begun.

Photo 5–9

The platform upon which Geoworld was to be built was a piece of tri-wall construction board, four by eight feet. The very first piece of architecture that arose was Tetrahedra Terrace, a series of connected tetrahedra. Then came the Cuban Embassy, an idea sparked from surveying atlases for possibilities. The Cuban Embassy was a large decorated cube with accompanying Cubans, who were represented by smaller cubes with personal characteristics. Many other structures were added to Geoworld; and when the project had been completed, every member of the class felt a deep sense of pride in the creative work of their peers.

Another project might be the construction of large space figures. For example a playhouse (or reading corner) may be a cube two meters on a side. A geodesic dome large enough for children to use could be constructed in the classroom and fitted with a light for reading. The possibilities for fun and adventure are unlimited.

Getting Ready to Measure

During this period of exploration and growth with geometric ideas, children generally develop to the extent that measuring can be successfully introduced. Piaget has identified two aspects of measuring that few children before seven-and-one-half to eight years old possess. He has indicated that measurement requires a combination of: (1) knowing that the measuring unit does not change in length as it is progressively moved along the distance being measured and (2) subdividing the distance to be measured into subunits of equal lengths (equivalent to the unit being used to measure). The first stage, conservation of length, volume, area, etc., is often reached by age seven. The coordination of both stages, necessary for the full ability to measure, occurs somewhat later. Thus, initial learning of measurement, (that is, before children are eight years old) should consist of informal work and much trial and error. Children cannot be taught how to measure. They learn through experimentation. Chapter 6 develops a sequence of activities for learning to apply number to space.

Photo 5–10

Extending Yourself

1.   Present one or two children (ages three- to five-years-old) with a set of materials such as building bricks, colored cubes, or logic blocks. Suggest they "make something" with the material. Without further verbal interaction, observe the range of creative play in which the children engage.

2.   Collect some drawings of several three-year-olds and compare them to the drawings of several six-year-olds. Note how the four basic relationships discussed early in this chapter are visually presented.

3.   Construct or otherwise obtain a geoboard. Explore relationships of perimeter (boundaries) and area of various figures. What is the perimeter and area of the largest and smallest square on the geoboard? What is the perimeter and area of the largest rectangle? The largest and smallest triangle? What problems arise when the precise perimeter is sought. How is one able to determine the area of irregularly shaped figures.

4.   Take a walk outside and keep a log of all geometric shapes observed in the environment. Look for examples of Euclidean figures, both two- and three-dimensional.

5.   Compare the activities suggested for children from five- to seven-years-old on the geoboard with those suggested for children ages seven to nine. In what ways do they differ? Why do they differ? How are the later geoboard activities related to the geometry activities for Parquetry Blocks? To pentominoes?

6.   Figure 5–26 illustrates three of twelve pentominoes. See if you can complete the remaining nine pentominoes. Collect milk cartons and determine how many can be cut to make pentominoes by actually cutting the cartons.

7.   It has been said that because of the way children view their world, geometry plays an important role in their mathematical education. Comment on this assertion.

Bibliography

Copeland, Richard. *Diagnostic and Learning Activities in Mathematics for Children.* New York: Macmillan Publishing Co., Inc., 1974.

——————. *How Children Learn Mathematics.* New York: Macmillan Publishing Co., Inc., 1974.

Cowan, Richard A. "Pentominoes for Fun Learning," *The Arithmetic Teacher,* Vol. 24, No. 3 (March, 1977), pp. 188–190.

Cruikshank, Douglas E. and McGovern, John. "Math Projects Build Skills," *Instructor,* Vol. LXXXVII, No. 3 (October, 1977), pp. 194–198.

Dienes, Z. P. and Golding, E. W. *Exploration of Space and Practical Measurement.* New York: Herder and Herder, 1966.

Dienes, Zoltan P. and Holt, Michael. *Zoo.* London: Longmen Group, Ltd., 1972.

Furth, Hans and Wachs, Harry. *Thinking Goes to School.* New York: Oxford University Press, 1975.

Morris, Janet P. "Investigating Symmetry in the Primary Grades," *The Arithmetic Teacher,* Vol. 24, No. 3 (March, 1977), pp. 188–190.

Nuffield Mathematics Project. *Environmental Geometry.* New York: John Wiley and Sons, Inc., 1969.

Piaget, Jean. "How Children Form Mathematical Concepts," *Scientific American* (November, 1953), pp. 202–206.

"Play Dome," *Sunset* (January, 1973), pp. 51–53.

Williams, Elizabeth and Shuard, Hilary. *Elementary Mathematics Today: A Resource for Teachers Grades 1–8.* Menlo Park, California: Addison-Wesley Publishing Company, 1976.

6

Children Applying Number to Space

Continuous and Discrete Measurement

Relationships between objects are defined by comparing properties of the objects. Some properties of objects can be quantified and described numerically. Measurement is a process of defining relationships between objects by comparing properties of the objects that are described numerically.

Most measuring activity involves a tool specially designed to assign a numeric value of so many units to some property possessed by an object. This type of measurement is called *discrete measurement*. However, some measuring tools do not assign a number but use some other means of making comparisons between the magnitude of measurable properties. Such measurement is called *continuous measurement*. An example of a continuous measuring tool for the property of length would be a piece of string. To determine whether the width of the teacher's desk or the width of a classroom window was greater, a piece of string could be stretched along the base of the window and cut to the appropriate length. The same string could then be stretched across the width of the desk, and a comparison of the width of the desk and window could be completed.

Young children need experience with two types of continuous (non-numeric) measuring techniques. The two techniques are direct and indirect comparison. *Direct comparison* is the process of comparing the magnitude of a measurable property of two objects by placing the objects in close proximity so that the comparison can be completed without the use of any intermediate object or measuring tool. In the example above, direct comparison would require that the desk be moved to the window so the difference in dimensions could be directly perceived without the use of a piece of string.

Indirect comparison is the process of using the string exactly as described in the example. The measurement is indirect, because the comparison between the desk and the window is performed by comparing each to a piece of string rather than by comparing the objects directly to one another.

Discrete measurement is a process by which some property of the object to be measured is divided into a number of discrete units of equal size. The measurement is performed by simply counting the number of units equivalent to the dimension being measured. For example, the width of a classroom window could be compared with the width of the teacher's desk by laying chalkboard erasers equivalent to the width of the window. The same procedure would then be performed on the teacher's desk. The object requiring fewer erasers would be the shorter object. It is necessary for children to acquire both conservation of length and number, before they are capable of understanding the concept of discrete measurement.

There are two aspects of discrete measurement that children should experience. In the example cited above, discrete measurement was performed using an arbitrary unit of measurement (erasers). Discrete measurement may also be performed using standard units, such as centimeters. Children should have experience first with arbitrary discrete units of their own invention and later with the standard discrete units.

In general, children should experience types of measuring activities according to the following sequence:

1. Continuous measurement with direct comparisons
2. Continuous measurement with indirect comparisons
3. Discrete measurement with arbitrary units
4. Discrete measurement with standard units

Children should move through these four stages at a rate suitable to their stage of cognitive development.

Another important influence on the children's rate of progress through the measuring sequence is the specific property that the children are measuring. There are six basic properties of objects that children should learn to measure. They are listed below in roughly the sequence that children acquire the ability to learn them:

1. Length
2. Area
3. Volume and capacity
4. Weight and mass
5. Time
6. Temperature

Other properties of objects will eventually be measured by children but are combinations of the above properties and are more complex. A few examples are:

1. Speed—a combination of length and time
2. Pressure—a combination of weight (mass) and area
3. Density—a combination of weight (mass) and volume

The precise sequence by which children acquire the ability to learn the various forms of measurement is not well known. However, it can be said with some confidence that continuous measurement experience should come prior to discrete experience, whereas measurement of length should precede area and volume.

Figure 6–1 is a graphic presentation of the general sequence that children may follow when learning measurement. Children should experience measurement concepts in roughly the sequence indicated by the line passing through the graph. For example, children should have their first experience with measurement by making continuous direct comparisons of length. Next, when children conserve length, they are ready to move on to continuous indirect measurement of length. After conservation of area is acquired, children can begin continuous measurement of area, first using direct and then indirect comparisons.

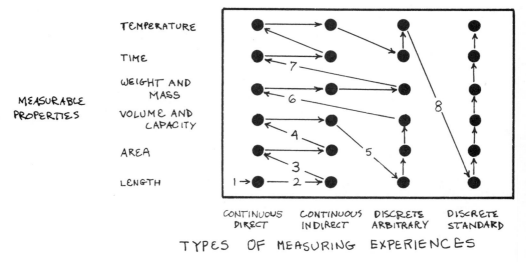

TYPES OF MEASURING EXPERIENCES

FIGURE 6–1

The numerals on the path in Figure 6–1 represent major developments in the children's cognitive abilities as listed below:

1. Premeasurement activity prior to conservation of length (ages 2 to 5).
2. Acquisition of conservation of length (ages 3 to 5).
3. Acquisition of conservation of area (ages 4 to 6).
4. Acquisition of conservation of volume and capacity (ages 4 to 7).
5. Acquisition of conservation of number (ages 4 to 7).
6. Acquisition of conservation of weight and mass (ages 6 to 9).
7. Concept of time and temperature as measurable properties (ages 7 to 10).
8. Readiness for standard units of measurement by virtue of: (1) having learned each aspect of measurement conceptually; and, (2) having invented arbitrary units of measurement for each measurable property (ages 6 to 10).

The remainder of this chapter will be a discussion of activities for each step of the sequence given in Figure 6–1.

Children Beginning Measurement

As shown in Figure 6–1, young children should begin measurement with the continuous direct measurement of length. The activities in this section will follow the sequence through the continuous indirect measurement of volume and capacity.

Continuous Direct Measurement of Length.

Each activity below engages children in directly comparing various lengths of objects.

1. Take two unused pencils of the same length. Line them up as in Figure 6–2a and ask the children if they are the same length. The children may wish to move them a little to convince themselves that they are the same length.

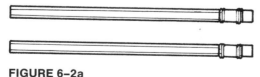

FIGURE 6–2a

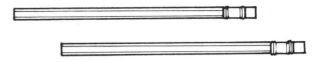

FIGURE 6–2b

Now move one pencil so the ends are no longer lined up as in Figure 6–2b. Ask the children if the pencils are still the same length, or if one is longer or shorter than the other. A belief that one pencil is now longer or shorter implies that the children do not yet conserve length. Children who do not conserve length should not be expected to give correct answers in the following activities.

2.   Provide children with ten or twelve common household objects that will stand up and are less than 40 cm tall; for example, a bottle, a brick, and a can. Spread the objects about so that no two of them are closer than one meter to one another. Ask the children: "Which object is the tallest? What is shortest? Which is taller than the can?"

Since the objects are not close together, it will be difficult to answer some of the questions accurately. Ask the children how they might find out for sure. With leading questions direct them to put the objects together in pairs to answer the questions accurately.

Photo 6–1

3.   Provide children with ten or twelve household items which do not stand on end; for example, toothbrush, pencil, magazine, and a serving spoon. Ask questions such as: "Which item is longer than the magazine? Which item is shorter than the toothbrush? Which item is the same length as the pencil?"

This activity requires the children to align one end of the pair of objects to be compared, before determining which is longer or shorter. In activity 2, the items were already aligned because they were sitting upright.

After children are proficient at answering questions such as the above, ask a few questions for which there are no correct answers. For example, assuming that the serving spoon is the longest item, ask: "Which item is longer than the serving spoon?"

Later, questions involving two conditions may be asked: "Which item is longer than the pencil but shorter than the magazine?"

4.   Using the children themselves as the objects to be measured, ask questions such as those given in activities 2 and 3. Have the children find who is tallest, shortest, who has the longest arm, the shortest foot, and so forth.

5.   Put two Cuisenaire rods of nearly the same length (yellow and dark green) into separate opaque bags. Let students take turns reaching into the bags and estimating which of the rods is longer by touch alone. Let the children conclude that to make the most accurate possible estimate, they must reach into both bags simultaneously and make a direct comparison.

Put a complete set of ten rods into one of the bags, and lay another set on the table in front of the children. Let children reach into the bag, grab a rod, and try to choose a rod from the table which is the same length. This activity can be played as a game with the student able to identify the most rods correctly being the winner.

6.   Choose an object in the classroom that is about as long as an average student is tall. Have the children compare their length, perhaps by lying on the floor, with the object and then separate themselves into three groups: longer than, shorter than, and same length as.

After children have done this activity with one or two different objects and their ability to conserve length has been demonstrated, choose an object that is inconvenient, such as the height of the chalkboard from the chalk tray to the top. Let the children try to invent a way to measure themselves against the board. Ask leading questions that direct them to the possibility of using indirect comparison.

Continuous Indirect Measurement of Length

Each activity below engages children in comparing lengths of objects using an intermediary object.

1.   Obtain a roll of craft or butcher paper at least 60 cm wide and about 30 m long. Cut the paper in lengths slightly longer than the children are tall. Have each child lie on a piece of the paper while another child draws a crayon outline of the child on the paper. Each of the children should sign his or her outline.

Now ask the children if they can determine who is taller and who is shorter than the chalkboard. Ask appropriate questions to assure the children understand they can draw accurate conclusions by comparing their outlines rather than their actual bodies.

Photo 6–2

Choose other relatively inaccessible objects in the classroom for comparison. Encourage children to draw conclusions such as: "Jimmy is taller than the closet is wide. Susan is shorter than the bottom half of the window."

2. Provide children with yarn, string or cash register tape, and scissors. Name an object in the room and let the children each estimate its length by cutting a length of their material they think will be as long as the object. Then let the children compare their estimate directly with the object.

3. With blocks or other similar material, let children build a tower at the front of the room on the floor and a tower at the back of the room on a table. Let the children try to determine which tower is taller. After all children have made their estimates, let them cut a string of the appropriate length to make an indirect comparison. Be certain that the children discuss which tower is *taller* rather than which tower is *higher*.

4. Give children some string and scissors and some length problems to solve in the classroom. For example, "Will the teacher's desk fit into the closet? Will the book case fit under the chalkboard?"

5. Give children some string and scissors and have them compare the lengths of parts of their body. Ask questions such as: "Who has the longest arm? Who has the shortest foot?" For some body part such as shoulder width, order the strings cut for each child and tape them to the chalkboard to form a graph (see Figure 6–3).

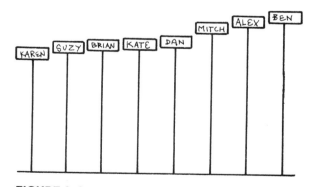

FIGURE 6–3

6. Tell the children that they are going to decide who has the tallest and the shortest person in their families. Send the children home with a length of yarn about 300 cm long. Tell them to measure the tallest person and the shortest person who live at their home and cut a piece of yarn the appropriate length.

When the children return to school, have them label the pieces of yarn with the family names. Clear an area of the floor for the "shortest person graph" and the "tallest person graph." Tape the pieces of yarn in order from shortest to tallest in each graph.

Discuss with the children the "tallest" and "shortest" graphs. The "shortest" graph will usually reveal the youngest member of each household. The "tallest" graph will usually imply the height that the student from that household may eventually attain. The order of heights from all of the households may also be compared with the order of heights of the children in the class. The orders will often be roughly the same.

7.   Choose one of the activities above, but give the children strips of crêpe paper to work with rather than yarn or cash register tape. Discuss the problems that occur if the measuring tool being used is capable of changing length.

Continuous Direct Measurement of Area

Each activity that follows involves children in directly comparing the area of various shapes.

Photo 6–3

1. Cut two pieces of green posterboard into squares of about 30 cm × 30 cm. Trim the squares so that they are identical in size. Obtain ten 2-centimeter blocks and a small plastic cow or a drawing of a cow. On one piece of posterboard, place five of the blocks in a tight group. On the other piece, place the other five blocks in some widely scattered pattern. Ask the children: "Which cow has the most grass to eat?" If necessary, tell a story about one farmer building one big house in his meadow and the other building five small houses scattered around the field.

The children should be able to determine with confidence that both cows have the same amount to eat no matter how the five blocks are distributed around the field. Any hesitancy on the part of the children implies that they do not conserve area. Children who do not conserve area should not be expected to give correct answers to the activities that follow.

2. Provide children with a set of flat, circular household objects: for example, lids from jars of various sizes, place mats, and so on. Ask the children which object has the most area. If they do not understand the concept of area, remind them of the story of the cow in activity 1, and ask which cow would have the most to eat if the circular objects were meadows.

Ask questions such as: "Which area is smallest? Which area is larger than the lid but smaller than the place mat? Place the areas in order from largest to smallest."

3. From posterboard or tagboard cut a set of figures as described below:

7 squares having sides of 5 cm, 6 cm, 7 cm, 8 cm, 9 cm, 10 cm, and 11 cm.

7 equilateral triangles having sides of 5 cm, 6 cm, 7 cm, 8 cm, 9 cm, 10 cm, and 11 cm.

7 diamonds having diagonals of 5 cm × 7 cm (as shown in Figure 6–4), 6 cm × 8 cm, 7 cm × 9 cm, 8 cm × 10 cm, 9 cm × 11 cm, 10 cm × 12 cm, and 11 cm × 13 cm.

Ask the children questions such as: "Which triangle has the greatest area? Which diamond has the greatest area? Which triangle has the smallest area? Which square has the smallest area? Put the diamonds in order from

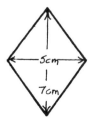

FIGURE 6–4

smallest to largest. Which figures have the most area, the squares or the triangles?" Many possible answers and much discussion likely will be generated when exploring these shapes.

When children begin to compare areas of different shapes, they will find it is easy to compare the areas of triangles with squares and diamonds with squares but very difficult to compare triangles with diamonds. This will imply the need for another way to compare the areas of objects.

4. Cut several pieces of posterboard into various irregularly curved shapes so they vary in size from about 20 cm² to about 120 cm² as shown in Figure 6–5.

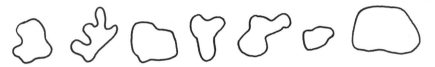

FIGURE 6–5

Place a paperclip on one of the medium-sized pieces and ask the children to find the "larger than" and the "smaller than" pieces. Change the paperclip to another piece and ask the same question. Ask the children to order the pieces from largest to smallest.

Some of these decisions may be difficult to make because of the irregularity of the pieces. Again, the need for another way to compare the areas of objects is implied.

5. Collect a few rectangular prisms from everyday household items such as a milk carton, a brick, and a small box. Trace the sides (faces) of the objects onto posterboard and cut them out. There should be five or six face-pieces for each object. Give the face-pieces and the objects to the children and let them try to determine which piece fits which side of which object.

The complexity of the activity can be increased by using several objects and cutting only one or two face-pieces for each object.

6. Provide children with 10 cm × 10 cm squares of posterboard. Ask them to trace the squares onto a piece of paper. If the children cannot trace accurately, provide them with a tracing of the square. Ask the children to cut the square into four irregular pieces. Ask: "Does the square still have the same area as before? Can you show that it does or does not?" Encourage the children to reassemble their squares inside of the traced square.

Repeat the activity with squares of different sizes and figures of different shapes such as triangles, circles, rectangles, and so on. Also, have the children cut figures of different sizes into a different number of pieces.

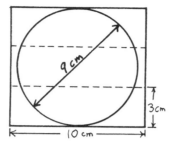

FIGURE 6–6

7. Provide the children with a rectangle with 30 cm × 3 cm dimensions and a circle with a 9 cm diameter. Ask the children to determine which has the larger area. The children will often conclude that the circle has the greater area. Tell the children to cut the rectangle into three equal parts (each 10 cm × 3 cm) and to reassemble the parts into another figure (a 10 cm × 9 cm rectangle). They can then compare the areas directly and will discover that the rectangle has the greater area (see Figure 6–6).

Photo 6–4

Continuous Indirect Measurement of Area

Each activity that follows involves children in comparing areas of various shapes using an intermediary shape.

1. Draw several rectangles of various shapes and sizes on the floor or chalkboard. The largest dimension of any particular piece should be about 30 cm. If possible, use different colors of chalk so the children can refer to them as the red rectangle or the blue rectangle, etc. Ask questions such as: "Which is larger? Which is smaller? Which is larger than the green rectangle and smaller than the orange rectangle?"

Because the children cannot move the rectangles around they must devise a way to make a comparison indirectly. It may be suggested that they cut a piece of paper the size of one of the rectangles to be compared and compare the paper to the other rectangle.

2. Present the children with problems similar to those in activity 1 above but involving different shapes. They should have experience with all of the standard straight-sided figures as well as circles, elipses, and irregular straight-sided and curved figures. The important principle for these activities is to provide shapes that cannot be moved, so that the children are forced to devise some way to do an indirect comparison. As in activity 7 of the previous set of activities, some of the shapes should be such that the children will be forced to cut and rearrange their measuring tool to determine which shape is larger.

3. Present the children with several problems involving immovable objects in the classroom. Ask: "Which is larger, one section of glass in the window or the top of a student's desk? Which is larger, the top of the filing cabinet or the top of the book case?"

For each problem, the children must cut a piece of butcher or craft paper the size of one of the objects and compare the paper to the other object. In most cases the objects will be of a different shape, requiring the children to cut and rearrange the paper to fit the second object.

4. Challenge the students with the problem of comparing the surface areas of their bodies. They might wrap each other up in bathroom tissue and then compare the amount of tissue used.

Continuous Direct Measurement of Volume

Each activity that follows involves children in directly comparing the volume of various objects or containers.

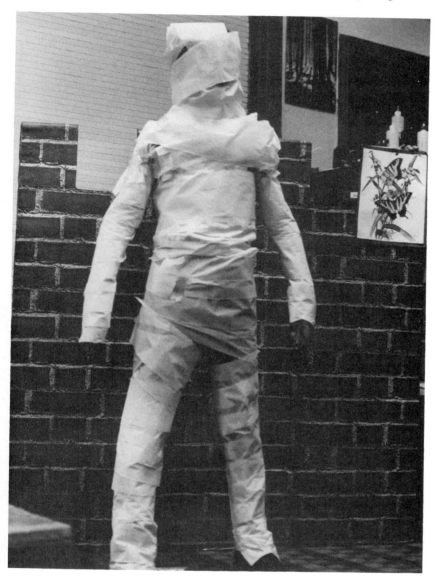

Photo 6-5

1. Take two glasses that are the same size and shape and one glass that is shaped differently. As in Figure 6-7a, pour an equal amount of liquid into the two glasses of the same size and shape. Ask the children if there is the same amount of liquid in the two glasses.

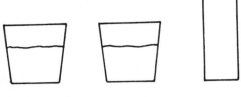

FIGURE 6–7a

The children may wish to add a bit more liquid to one glass before an agreement is made that the glasses have the same amount of liquid. As in Figure 6–7b, pour the liquid from one of the two glasses into the third glass of a different shape.

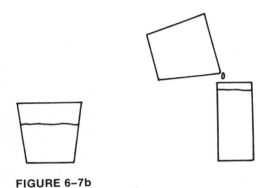

FIGURE 6–7b

Ask the children if the glasses still have the same amount of liquid, or if one has more or less than the other. A belief that the amount of liquid has changed indicates that the children do not yet conserve volume. Children who do not conserve volume should not be expected to give correct answers in the following activities.

2. Play the "Large to a mouse but small to an elephant" game. Ask questions such as: "What objects in the room seem small to you but would seem large to a mouse? What objects in the room would seem small to a mouse? What objects that seem large to you would seem small to an elephant? Are there any objects that seem small to you that would seem large to an elephant? Why?" Continue asking questions until all of the possibilities have been exhausted. Then name an object and ask: "Is it large or small to you? To an elephant? To a mouse?"

3. Provide children with several boxes of various sizes and colors, so that the smaller boxes will fit inside the larger boxes. Ask: "Which boxes are smaller than the red box? Which are larger? Which is the largest box of all? Which boxes are larger than the blue box, but smaller than the red box?"

Continue questioning until all possible relationships are exhausted. After the children have acquired some proficiency with volume, you may include some boxes that are roughly the same size, but different shapes. The children will be unable to determine which box is larger by direct comparison, thus creating the need for the next step in the learning sequence—indirect comparison of volume.

4. Make up six to eight balls of clay of various amounts. Include graduated amounts from approximately 15 cm³ up to 100 cm³. After making up the sequence of balls, change the shape of each ball drastically, so that each wad of clay looks completely different; for example, make a flat disc, a long tube, a cup, and a doughnut. Ask the children the usual questions: "Which has more clay, the tube or the cup? Which has less clay, the disc or the doughnut? Which has more clay than the doughnut, but less than the cup?"

Ask the children how they can show their answers are correct, and encourage them to reshape the clay into similar shapes—not necessarily balls.

Continuous Indirect Measurement of Volume

Each activity that follows involves children in comparing the volume of various shapes and containers using an intermediary volume.

1. Provide children with a supply of containers of all shapes and sizes. Obtain a small child's swimming pool or other appropriate container in which water can be poured with minimal mess. Ask the children: "Which container holds the most water? Which container holds the least water? Which containers hold more than the bleach jug? Which containers hold more than the shampoo tube, but less than the red drinking cup?"

Allow children to experiment freely and to pour the water back and forth to prove their conclusions. Encourage children to estimate their answers before checking them. After the children have become rather proficient at using the water to compare volumes of unlike containers, ask them to order the containers from "contains least" to "contains most."

2. Provide several clear containers having widely varying shapes and volumes. Using some container smaller than any container in the clear group, pour exactly the same amount of water into each of the clear containers. Discuss with the children the disparity between the heights of the water in the containers. Ask the children whether or not they can order the containers according to their capacity by looking at the water levels.

3. Place four or five containers around the room and tell the children to determine which is largest and smallest and to order the containers without moving them. Be certain to place the containers in positions where water spills will do no harm.

4. Make several wads of clay of various sizes and obtain a clear straight-sided container into which each of the wads will easily fit. Partially fill the container with water, so that the largest wad of clay can be completely submerged without causing the water to flow over the top. Attach each wad of clay to a stick or stiff wire and submerge it into the water. Let the children observe and discover that the water level rises when the clay is submerged. Ask the children if they can use this fact to determine which wad of clay has the greatest volume.

After the children can confidently compare volumes using the water displacement technique, ask them to bring a rock of irregular shape from home. Let them compare the volumes of their rocks.

Children Measure with Arbitrary Units

When children have acquired conservation of number, they should be introduced to discrete arbitrary measures. They may continue with continuous measures of weight, mass, time, and temperature when these concepts are also developed.

Discrete Arbitrary Measurement of Length

Each activity that follows engages children in determining the length of various curves and objects by repeatedly using an arbitrary length.

1. Draw two curves of approximately three meters length on the floor with chalk. Stand one child at each end of one of the curves, then ask the children: "How many of you can stand comfortably on the curve between the two children on the ends?" After they give estimates, let them measure the curve by standing on it shoulder to shoulder, so that each of the children lightly touches the ones next to him or her.

Have the children measure the next curve in the same fashion. After they determine the number of children for both curves, ask: "Which of the curves do you think is longer? Why?"

2. Provide the children with ten each of several objects having a distinct length; for example, pencils, toothpicks, tongue depressors, paperclips. Have the children lay each set of ten objects end-to-end on the floor. Ask: "Which set of objects makes the longest line?" Why? Put the lines of objects in order according to length. Put one of each object in order with the rest of the objects. How are the two orders alike?"

3. Divide the class into five or six groups; give each group a sufficient number of one of the measuring objects from activity 2. Assign each group to measure some object in the classroom, such as the width of a window, height of a table, or length of a book case. Record the results of each measurement on the board; for example, the window equals 7 pencils, a table equals 27 paper-clips, and so forth. Ask the children if they can determine which object is the longest by using the measurements they have found. Be certain that they understand that if two objects are to be compared by this type of measurement, then both objects must be measured by the *same unit*.

4. Divide the children into groups of four to six, and give each group a supply of tongue depressors. Have each group measure several objects such as those in activity 3. Because it is unlikely that most objects will be exactly an even number of tongue depressors in length, have the children record their results in the form, "greater than six but less than seven."

Record the results of the measurements in a chart as shown in Figure 6–8, p. 202.

Photo 6–6

Window	more than _____ but less than _____
Teacher's Desk	more than _____ but less than _____
Student Desk	more than _____ but less than _____
Chalkboard	more than _____ but less than _____
Door	more than _____ but less than _____

FIGURE 6–8

Discuss whether the students can determine from their measurements if the window or the student desk is longer. Ask the children to measure the window and desk again, but this time have them measure it with a paperclip chain. Discuss the fact that a smaller unit of measurement will result in a more precise measurement.

5.   Divide the class into groups of four to six children, and provide each group with a set of Cuisenaire rods. Have the children measure several objects in the room, but instruct them to use as few rods as possible for each measurement. This requirement will result in measurements of decimeters and centimeters. The children should record their measurements in the form "three orange rods and one black rod," "five orange rods and one red rod," and so on.

6.   After the children have mastered the measuring process with Cuisenaire rods, tell them that they are going to make a measuring tape that will make their measuring tasks easier. Provide each child with a piece of cash register tape approximately one meter long and a set of rods. Have the children lay the orange rods along the tape, and make a line across the tape at the end of each orange rod. Then have the children mark the length of each other rod between the orange rod marks as shown in Figure 6–9, p. 203.

If the children are familiar with the Cuisenaire rods, they will realize that each mark on the tape is equivalent to another white rod. Suggest to them that they might record their measurements as orange and white rods only. For example, the window might be greater than 7 orange and 3 white, but less than 7 orange and 4 white.

Again, divide the children into groups, and have them measure several classroom objects. This time they will find that all groups will arrive at the same answers, because they are all using identical measuring instruments.

7.   Have the children bring to class several toy cars that roll easily. With some books and stiff cardboard, make a ramp for the cars to roll down. Mark a starting line on the ramp. Let the children roll their cars down the ramp, one at a time, and measure the distance rolled with their tapes from activity 6. Record

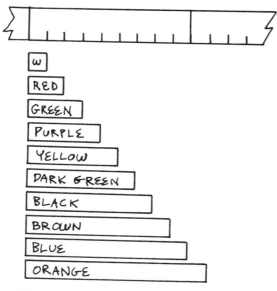

FIGURE 6-9

the distances in tabular form on the chalkboard or on a prepared mimeographed sheet. Ask questions such as: "Which car rolled the greatest distance? Which car rolled the shortest distance? Which car rolled farther than
the blue car, but not as far as the red car?"

Discrete Arbitrary Measurement of Area

Each activity that follows engages children in determining the area bounded
by closed curves by repeatedly using an arbitrary unit of area.

1. Draw two arbitrary closed curves on the floor, each enclosing an area
of approximately 2 m². See how many children can stand within each region.
Both feet must be inside the curve, and all children must be standing comfortably upright. Ask the children if they can decide which curve encloses the
most floor in it (area). Draw other curves, letting the children stand within the
curves and recording the number. Ask questions such as: "Which curve encloses the least area? Which curve encloses more area than curve A, but less
area than curve C?"

Draw two curves of roughly the same size. Let the children try to decide
which is larger. Suggest that a better way to determine the area of curves is
needed.

2. Provide the children with several rectangular objects such as books,
sheets of paper, and index cards. Also give them a supply of blocks of equal
size. Tell them to determine how many blocks it takes to cover a book, a sheet

Photo 6–7

of paper, and so forth. Have the children record the results in tabular form. They may make a "physical table" with the blocks themselves, as shown in Photo 6–7.

Ask questions such as: "Which object has the most area? Which has a greater area than the book, but less than the paper?"

3. Using a set of textbooks of equal size, let the children compare the areas of several large objects in the room, such as the top of the book case and the top of a student desk. Again, the children should record their results in tabular form and answer comparative questions. Some of the objects may be too large to be completely covered by the set of books available. Let the children try to solve the problem by laying all of the books on the object to be measured, counting them, and then moving part of them to cover the rest of the object.

4. Provide the children with a supply of blocks of uniform size (approximately 5 cm on a side). Have the children make patterns with the blocks and trace around them on paper. Let other children try to guess how many blocks were used for the pattern and check their guesses by placing the blocks on the pattern (see Figure 6–10).

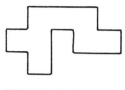

FIGURE 6–10

5. Prepare mimeographed sheets of 2 cm $\times$ 2 cm² grids. Let the children take any objects that will fit on the grids and trace their outlines. Tell the children to count the total number of squares that the object touches partially plus the total number of squares that are completely inside the trace of the object.

The tape dispenser in Figure 6–11 has four squares completely inside, plus it touches another seventeen squares. Therefore, the area of the dispenser is greater than four squares, but less than 21 squares.

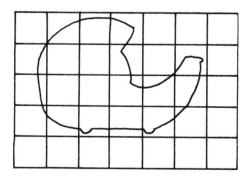

FIGURE 6–11

After the children have measured several objects in this fashion and recorded the result in tabular form, let them try to decide which is larger and which is smaller. Since it will be difficult or impossible to compare many of the objects using this method, let the children try the same objects with grid sheets of one cm². If it is still not possible to determine some of the comparisons, let the children suggest that a grid having smaller squares should be used.

6. Provide the children with geoboards (see Appendix) and rubber bands. Encourage them to make figures of various shapes and sizes and count the number of squares in the shapes. Since all shapes on the geoboard have straight sides that terminate at intersections of the lines of the geoboard grid, it is always possible to find the area of a shape to the nearest square. That is, all areas can be determined as greater than x squares, but less than y squares, where x and y are consecutive whole numbers.

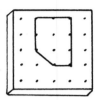

FIGURE 6–12

Discrete Arbitrary Measurement of Volume

Each activity that follows engages children in determining the volume of various containers, by repeatedly using an arbitrary unit of volume.

1. Provide the children with several jars or other containers having volumes from 500 milliliters to two liters. Also provide a supply of 100 milliliter paper cups. Ask the children to find the largest and smallest container, by finding how many cups of water each container will hold. The children should express the result of their measurements in a table. Some of the containers may be more or less than an exact number of cups. In such cases, the children should express the volume as "more than x cups but less than y cups."

2. Provide about 20 containers holding from 50 milliliters to about 1500 milliliters and having a wide variety of shapes and configurations. Let the children in pairs play the following game:

Brandon takes containers A (small) and B (large) from the stack and measures to see how many of the A's are required to fill the B. Then the A container is returned to the stack, and Brandon gives container B to Craig and tells Craig how many A's were required to fill it; for example, four. Craig, who does not know the identity of container A, must experiment with containers from the stack until he correctly identifies the container that is one fourth of container B.

The children can keep score by recording the number of containers guessed before finding the correct one. The person who makes the fewest incorrect guesses wins.

3. Have the students bring a number of small boxes to class ranging from jewelry-size to shoe-box-size. Provide them with a supply of 2 cm cubes, and let them determine how many cubes fit into each of the boxes. They should record their results in tabular form and should respond correctly to comparative questions regarding the volumes of the boxes.

4. Provide the children with clear containers having straight sides and capable of holding at least one liter of water. Ordinary quart jars would be sufficient. Using small cups of approximately 100 ml volume, have the children pour one cup of water into the jar, and mark on the side of the jar the level of the water. Then pour a second cupful and mark the new level, and so forth until the jar is full and there is a mark for each cup of water poured.

Now the jars can be used to measure the volume of other objects. Provide the children with a number of other smaller containers, and allow the children to measure the volume by filling the small container with water and pouring it into the measuring jar. The children should record their results in tabular form and should correctly answer questions regarding the relative volumes of the containers.

5. Provide the children with a jar having arbitrary volumetric units marked on it, as in activity 4. Also obtain several irregularly shaped objects such as stones that can be submerged in the container. Fill the container about half full, so that the water line exactly matches one of the lines on the side. Let the children submerge the objects in the water, and record the rise in water level for each. They should record their results in a table and should correctly answer questions regarding the relative volume of the submerged objects.

Continuous Direct Measurement of Mass

Each activity that follows engages children in comparing the mass of various objects.

1. Form two balls of clay that have the same mass. Ask the children if the two are the same. The children may wish to add or take away a little clay from one ball before it is determined that the balls have the same mass. Chop one ball up in small pieces. Give them to the children, and ask if they still have the same mass, or if one is heavier or lighter than the other. A response that one is heavier or lighter indicates that the children do not conserve mass and should not be expected to give correct responses to the following activities.

2. Provide a set of ten or twelve relatively small common household objects having masses of 100 g to 1 kg; for example, a pencil, spoon, book, transitor radio, and cup. Let children compare masses of objects by holding an

object in each hand and deciding whether one object is greater than, less than, or equal to the other object. As decisions are made, the children should lay the objects in a line from lightest to heaviest. The children should continually check and recheck pairs of objects until they are satisfied that their order is correct. They should record the order they have decided on.

The teacher should choose objects, so that some of greater size have lesser mass. For example, a small empty box would appear larger than a lead sinker, but the sinker would have a greater mass. Careful choice of objects will prevent the children from equating the property of size with the property of mass.

3.   Construct a crude beam balance from a 10 cm × 80 cm board and a triangular block (see Figure 6–13).

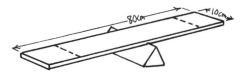

FIGURE 6–13

Let the children compare the masses of the same objects from activity 1. The children should establish an order for the objects by checking and rechecking their masses using the beam balance. Be certain that the children place the objects to be weighed inside the 10 cm × 10 cm area at each end of the beam. The distance of the object from the center of the beam may cause the object to appear heavier or lighter. After the children have established the order of the objects by mass, they should compare the results with the results of activity 2. If there are discrepancies, discuss the relative accuracy of the two methods and possible reasons why differences occurred.

4.   Obtain a pan balance and let the children solve the same problem as in activities 2 and 3. Again, have the children record the results, and compare with activities 2 and 3. The results for this activity will be the most accurate, since the pan balance keeps the distance from the fulcrum to the object being weighed constant.

5.   Fill one-half pint milk cartons labeled A to J with varying amounts (50, 100, 150, . . . 500 g) of plaster of paris or clay. Seal the cartons. Let the children order the objects by mass using the hand comparison, the beam balance, and the pan balance. Since the objects all look alike, the children must order the objects by mass alone, rather than by size or appearance.

6.   For each object in activity 2, let the children use a pan balance to find an amount of clay equal in mass to each object. After each amount of clay is found, roll each mass of clay into a round ball. Let the children speculate re-

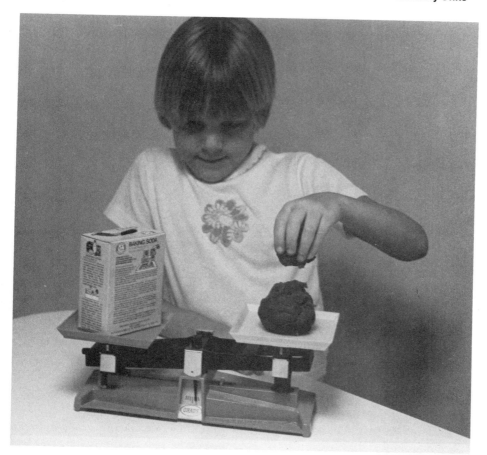

Photo 6–8

garding the relative size of the balls of clay and the mass of the objects. Ask leading questions to direct the children to the generalization that: For a given continuous material, such as clay, volume and mass are directly proportional.

7. Fix two special balls of clay. Make one large ball with a ping pong ball or other light object concealed inside it, Make another smaller, but heavier, ball by concealing a lead sinker inside it.

Ask the children which ball has the greater mass. Place the balls on the pan balance, and demonstrate that the smaller ball has the greater mass. Ask the children to explain why. If the children have previously completed activity 6 with understanding, they will suspect a trick since the result defies a known relationship between mass and volume.

8. Make a beam balance from a large board, approximately two meters long and 30 cm wide, and a brick. Use the brick as the fulcrum. Let the children order themselves by mass by standing on opposite ends of the board. Children may also stand two at a time on each end to compare themselves in pairs. How many children does it take to equal the mass of one teacher?

Continuous Indirect Measurement of Mass

Each activity that follows involves children in comparing the mass of various objects using an intermediary mass.

1. Collect two sets of 8 to 10 common household objects having masses varying between 50 g and 1 kg. Place one set of objects and a pan balance at the front of the room and the other set of objects and a pan balance at the back of the room. Tell the children that they are to find the heaviest and lightest objects in either of the two groups, but that they may not compare the objects at the front of the room directly with the objects at the back of the room.

You may wish to suggest that the children compare an object in the first set at the front of the room with a ball of clay, then carry the clay to the back of the room to compare with an object in the second set. Let the children continue to make comparisons until the heaviest and lightest objects from either set are found. If the children want a more difficult challenge, tell them to order all 20 objects (10 at the front of the room and 10 at the back) into a single order by mass, but without making direct comparisons between any objects from the separate sets.

2. Using the two sets of objects from activity 1, or similar sets, let the children compare masses of the objects using the volume comparison method. Provide the children with the two sets of objects, two pan balances, and two identical clear containers suitable for holding water. Place an object from the front of the room on one side of the pan balance and the container on the other side. Fill the container with water until balance is achieved. At the back of the room repeat the process with another object and the other container. Bring the containers together and compare the volume of water in each. Ask leading questions until the children understand that the volume of water is proportionate to the mass of the object against which it was weighed. Let the children order the objects from the two sets using this method.

This activity is a prelude to the later introduction of the standard mass unit, the kilogram.

Discrete Arbitrary Measurement of Mass

Each activity that follows engages children in determining the mass of various objects by repeatedly using an arbitrary mass.

1. Collect several sets of 50 uniform objects having masses from 10 g to 50 g; for example, a set of lead sinkers, set of washers, set of 2 cm blocks, set

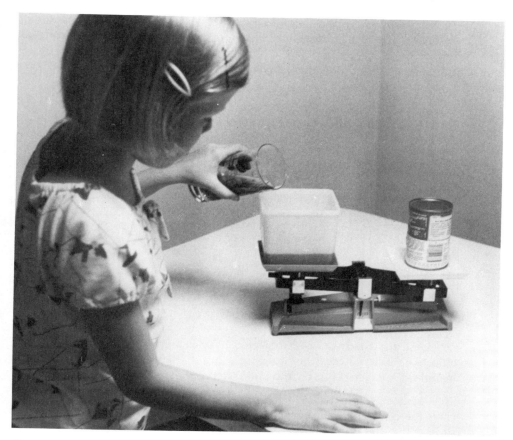

Photo 6–9

of bolts, and so forth. Divide the set of washers into subsets of one, two, three, . . . nine washers. With the beam balance, have the children compare the mass of the set of one washer with the mass of the set of two washers. Compare the other sets of washers until they are ordered by mass. The children should quickly note that as the number of washers increases, the mass of the set increases.

Perform the same activity with the set of lead sinkers and the other sets until it is certain that the children have generalized the relationship between the number of objects and the mass of the set of objects.

2.   Using ten objects from each of the sets of uniform objects in activity 1, let the students compare the relative weights of sets of ten. That is: "Which has the greater mass, ten washers or ten lead sinkers?" Let the children compare and order the sets of ten objects.

Perform the same activity, but with sets of five objects. Note that the order is the same for five objects, as it was for ten. The children should conclude that the same amount of different objects have different masses, and that if object A has a greater mass than object B, then ten of object A will have a greater mass than ten of object B.

3.   Using the sets of uniform objects from activities 1 and 2, and a pan balance, have the children weigh a number of objects from the classroom using washers, sinkers, and 2 cm blocks, and record the results in a table as shown in Figure 6–14.

	Washers	Sinkers	2 cm Blocks
Stapler	> 20 but < 21	> 15 but < 16	> 30 but < 31
Book	> 30 but < 31	> 22 but < 23	> 45 but < 46
Ball	> 5 but < 6	> 3 but < 4	> 7 but < 8
Toy car	> 3 but < 4	> 2 but < 3	> 4 but < 5
Paste bottle	> 5 but < 6	> 3 but < 4	> 7 but < 8

FIGURE 6–14

Ask the children which object is heavier and which is lighter. Can the objects be ordered according to mass using the table? Do the measurements made with washers, sinkers, and blocks agree with one another?

Notice that the ball and the paste bottle appear to have the same mass. Have the children find the mass of the objects again, but this time use paperclips as the unit of mass. Many more paperclips will be needed since the paperclip is a much lighter unit than the washer, sinker, or block.

The children should begin to recognize the principle of precision from their study of arbitrary units of length, area, and volume. To make a more precise measurement, decrease the size of the unit of measurement being used.

Continuous Direct Measurement of Time

Each activity that follows engages children in determining time by directly comparing events.

1.   Stage several events in the classroom, and ask the children to decide which event takes the longest time and which takes the shortest. Examples of possible events are:

1.   Hopping around the room on one foot.
2.   Writing one's name on the chalkboard ten times.

3.   Reciting "Mary had a little lamb."
4.   Tying two shoes.
5.   Making a stack of ten blocks.
6.   Unbuttoning and rebuttoning shirt.
7.   Stringing a set of ten beads.

At first, stage the events separately, and ask the children to order them according to duration. There will be much disagreement. Next, take two events and begin them at the same time. It will be easy to see which takes longest. Using this method, let the children order the events from least to greatest duration.

2.   Collect a set of objects that will fall differently, for example, balloon full of air, marble, leaf, large clay ball, small clay ball, sheet of paper, wad of paper, feather, pencil, and so on. Let one student stand in a chair or on a desk and drop two objects simultaneously. Have other students determine which object takes the greatest time duration to reach the floor.

Make an ordered list of the objects from greatest time duration to least time duration. Discuss the apparent incongruities, such as the balloon, that although much larger, takes longer to reach the ground than the marble.

3.   Obtain equal-sized containers of about 500 ml of tap water, hot tap water, and cold water from the drinking fountain. Drop an ice cube in each and compare the time duration required for each ice cube to completely melt.

Repeat the above activity but include both cold, cool, and hot fresh water and very salty cold, cool, and hot water. Dissolve 30 to 40 ml of salt in 400 ml of water. The children will find that the time duration for ice to melt in salt water will be different from the duration for fresh water. Make a table and record the water causing the longest duration, the shortest duration, and the order of time durations required for the other states of water.

4.   Obtain different temperatures of water as in activity 3, but in different volumes; for example, 100 ml of hot water and 200 ml of hot water. Compare the time duration for melting ice cubes in different amounts of water.

5.   Obtain the three different temperatures of water as in activity 3, and compare the time duration required to dissolve an Alka Seltzer tablet in each.

Continuous Indirect Measurement of Time

Each activity that follows involves children in comparing times of different events using an intermediary unit of time.

Photo 6–10

1. Have the children make water clocks by punching holes in the bottom of paper cups. Make several clocks with holes of different sizes. Compare the time duration for water to drain from each of the cups. Use one cup as a standard, and classify the others into groups such as "takes more time to drain than cup A," "takes less time to drain than cup A," or "takes the same time to drain as cup A." Put the cups in order according to the length of time it takes for them to drain.

2. After the water clocks have been ordered, stage several classroom events such as those listed in activity 1 of the previous section. Stage the events separately, and compare each to the water clocks. Try to determine which event takes the longest time by using the water clocks. Which event takes the shortest time? Try to order the events using the water clocks.

3. After the children have become proficient at timing events with the water clock, make up a set of very accurate water clocks by carefully punching holes of specific size in the bottoms of a set of paper cups. Time the cups in advance to assure that you have cups that have durations of approximately one minute, three minutes, five minutes, 10 minutes, 15 minutes, 20 minutes, and 30 minutes. Arbitrarily label the cups A, B, C, . . . G.

Use the cups to time daily classroom activities such as story time, clean up time, recess, and so forth. Increase the childrens' proficiency at certain events by racing against the cups. For example: "Everyone cleared off their desk yesterday before cup C ran out. Do you suppose you can beat cup B today?"

4. Let the children bring several windup toys to class. Wind them, and let them run down at separate times, and let the children estimate which toys take the longest and shortest times.

After they have made their estimates, let them compare the toys with the water clocks. Order the toys according to time duration.

Continuous Direct Measurement of Temperature

Each of the activities that follow, engage the children in directly comparing the temperature of various objects.

1. Have students place their hands on their cheeks to feel how warm or cool they are. Then have students briskly rub hands together for about 30 seconds, and put them against their cheeks again. Ask them if their hands were warmer or cooler, after they were rubbed together.

2. Arbitrarily label several spots in the room that have different temperatures for the students to touch; for example, radiator, window glass, metal shelf, spot in the sunlight, spot in the shade, and so forth. Let the students try to determine which spot has more heat and which spot has less heat.

Let the students make statements such as, "The metal shelf has more heat than the window glass but less heat than the radiator." Let the children attempt to order the spots by the amount of heat they feel. Point out to the children that the things that have more heat than their hand feel warm or hot to them, and the things that have less heat than their hand feel cool or cold to them.

3. Draw two glasses of water from the tap, and let the children feel them to see that they are both cool. Set one of them in the shade and one in the sun. Thirty minutes later, let the children again compare them. Discuss why one glass is now warm whereas the other is still cool.

Photo 6–11

4.   Get several pieces of material that are known to transmit heat at different rates; for example, metal, glass, rock, wood, wool, and so forth. Let the children feel the objects and try to order them according to which seems to have more heat and which seems to have less heat.

Teachers should realize that the materials are actually the same temperature but seem to have more or less heat because they transmit heat away from the body at different rates. This concept, however, is too difficult for children. The temporary misconception engendered by this activity will correct itself, as the children begin to understand temperature as a measurement of molecular activity.

5.   Let two student judges stand in front of the room. Have five other students file past and lay their hands on the judges' cheeks. The judges try to decide who has the hand with the most heat and who has the hand with the least heat.

Let the judges turn their backs. Have three students change the temperature of their hands by waving them in the air, rubbing them together, or breathing into them. Have each student immediately touch the judges' cheeks, and let the judges try to decide who was waving their hands, who was breathing into them, and who was rubbing them together.

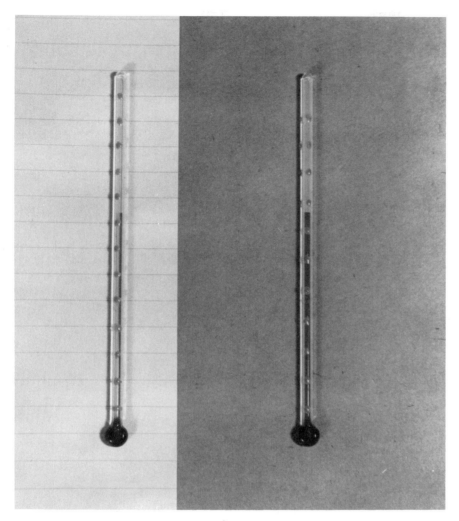

Photo 6–12

*Continuous Indirect Measurement of
Temperature*

Each activity that follows involves children in comparing the temperature of
various materials by using an intermediary source.

    1. Remove the glass bulbs from several inexpensive thermometers so
that no scale is visible to the children. If the bulbs are removed carefully, they
can be replaced and used as standard thermometers at a later time.

Let the children discover that the bulbs indicate more heat when the mercury column gets longer and less heat when it gets shorter. Let the children find the warmest and coldest spots in the classroom. Remind the children that they must lay the bulb down to make a reading. Holding it in their hand will only record the amount of heat in their hand.

2.   By mixing hot and cold water, fill several cups with water of different temperatures. Let the children order the cups from least heat to most heat by using the thermometer bulbs.

3.   From activity 4 of the previous section, lay the metal, glass, rock, wood, wool, and so on in order on a table. Let the children try to verify the results of their experiment in activity 4 of the previous section by using the thermometer bulbs. Discuss why the thermometer bulbs all give the same reading, in spite of the fact that the materials feel as if they have different amounts of heat.

4.   Lay two bulbs side-by-side on the window sill so that one is in the sun and one is in the shade. Discuss the drastic difference in the mercury column, in spite of the fact that the two bulbs are only centimeters apart on the same sill. Discuss why days are warmer than nights and why cloudy days are cooler than sunny ones.

Discrete Arbitrary Measurement of Time

Each activity that follows involves children in comparing various lengths of time using an arbitrary device of fixed time duration.

1.   Construct several pendulums using string and masses, such as a lead sinker or a stack of three or four washers. Let the children play with the pendulums, comparing the rate of the swing as the length of string is lengthened or shortened.

2.   Decide upon an arbitrary length for all pendulums in the classroom to standardize results of experiments. Divide the class into groups, and give each group a standard pendulum and two or three water clocks. Let each group find the number of swings of the pendulum required for each water clock to run out. List the results in a table on the chalkboard.

Water Clock	Pendulum Swings
A	16
B	27
C	48
D	12
E	32
•	•
•	•
•	•

From the table, let the class decide which water clocks take the longest time and which take the shortest.

3. Make a large pendulum of heavy string about two meters long and a mass of about 500 g. Hang the pendulum from the ceiling or other high place so that the entire class can see it. Time classroom events using the pendulum and counting swings out loud.

"Can you clear your desk and be ready to leave for lunch before 50 swings of the pendulum? Can you complete a two-digit multiplication problem before 20 swings of the pendulum?"

Discrete Arbitrary Measurement of Temperature

Each activity that follows involves children in determining the temperature of various materials using an arbitrary unit of temperature.

1. Using thermometer bulbs as in activity 1 in the previous section, and fingernail polish, have the children make arbitrary scales for their mercury columns. Scales also can be marked by laying the bulb on a sheet of notebook paper and making a mark on the column for each line on the paper. Be sure to align the bottom of the bulb exactly on one of the lines before making the marks.

Prepare several cups of water of different temperatures by mixing hot and cold water from the sink. Divide the children into groups, and give each group a cup of water and a marked thermometer bulb. Have each group measure the temperature of their water, and record the result in a table on the chalkboard.

Cup	Lines on the Bulb
A	> 3 but < 4
B	> 6 but < 7
C	> 3 but < 4
•	•
•	•
•	•

Let the children try to order the cups from least heat to most heat by reading the results on the table. Some of the cups may fall between the same two lines as for A and C above. From previous experience with length, area, mass, and volume measurement, the children should realize that a smaller unit of measurement (marks on the bulb closer together) is necessary to find whether A or C has the most heat.

2. Using the marked bulbs from activity 1, let the children find and record the indoor and outdoor temperatures for each hour of the day, and record

Photo 6-13

the results in graphic form. Be certain that the outside thermometer bulb is not in direct sunlight and is exposed to the outside air. Let the children discuss their findings on the graph.

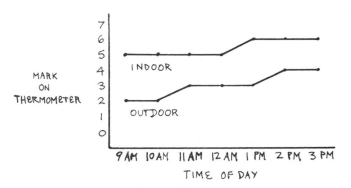

FIGURE 6-15

Children Using Standard Measurement

When children have mastered the earlier measurement concepts, they are ready for measurement with standard systems. The standard system chosen is not crucial. It could be the English or the metric system. The metric system is used here because of its reliance on base ten and of the necessity for children to know this system in the world in which they will live.

Discrete Standard Measurement of Length

Each activity that follows involves children in determining various lengths using standard units of length.

1. Provide the children with cash register tape and an abundance of centimeter cubes. Let the children make centimeter tapes with which to measure. Tell the children that the units they are using are named *centimeters*. Provide them with a number of objects to measure, and have them record their measurements in the form > 12 cm but < 13 cm. Choose the objects to be measured so that several of them will have the same measurement to the nearest centimeter.

2. Introduce the 30-cm ruler to the children, and have them compare their centimeter tapes to the ruler. Point out the smaller marks on the ruler, and tell the children that these units are called millimeters. Let the children count the number of millimeters in a centimeter. At this point, it is unnecessary for the children to know that the centimeter and millimeter are each based upon the meter. Introduce the two units as if they were just another arbitrary unit for measuring length.

Let the children measure the objects from activity 1 that had the same measurement to the nearest centimeter. They should record the result in tabular form.

Object	Measurement
A	12 cm and 3 mm
B	15 cm and 7 mm
C	12 cm and 4 mm
•	•
•	•
•	•

Ask the children if they can now determine from the table which object is longer and which is shorter.

3.   Divide the class into groups and provide each group with a meter stick. Explain that the whole stick is a unit of measurement called a meter. The children will also notice that the familiar centimeters and millimeters are marked on the stick.

Provide each group of children with a number of objects to measure which are more than one meter long. They should record their answers in the form: 2 meters and 21 centimeters and 7 millimeters.

Let the groups compare their results with one another, and discuss why some measurements of the same objects were different.

4.   Divide the class into small groups and tell them that they are going to use measurement to find some secret objects. Provide each group with a set of measurements of objects in the classroom; for example, object A is 91 cm and 7 mm tall, and 42 cm and 3 mm wide; object B is 11 cm and 4 mm long; and so forth. Each group must measure objects in the classroom until they identify the objects described by the measurements.

Photo 6–14

5. Conceal a hidden "treasure" on the playground and, using a meter stick or trundle wheel, mark off several paths to the treasure from various points on the playground. Use descriptions such as: left, 5 m; right, 12 m; right, 3 m; left, 19 m; and so on or use north, south, east, and west, if the students understand the terms.

Divide the class into groups of four or five, and take each group to one of the starting points on the playground. Provide each group with two meter sticks and a set of directions. The group that correctly follows their directions first will find the treasure.

6. Set up a "Metric Olympics." Let the children measure the track, high jump, and so on.

Discrete Standard Measurement of Area

Each activity that follows involves children in determining the area of various shapes using standard units of area.

1. Provide the children with centimeter grid paper and 30 cm rulers. Tell the children to measure one of the squares on the grid to see how long and wide it is. Explain that each square is called one square centimeter.

Provide the children with a set of irregularly shaped cutouts from index cards, and tell them to trace an image of the cutouts onto the grid paper. The children should count the number of centimeter squares inside the image plus the total number that the image touches without completely containing. They should record their answers in the form: Image A is greater than 8 cm² but less than 13 cm².

2. Provide the children with cm² grid paper, a 30 cm ruler, and a set of rectangles having integer centimeter measurements on each side; for example, 2 cm by 3 cm, 6 cm by 8 cm and 5 cm by 3 cm. Tell the children to measure the length and width of each rectangle and record their answer. Next, multiply the length times the width, and record their answer. Next, trace the rectangle on the grid paper, and find the number of cm² in each rectangle. Record all of the data in a table.

Rectangle	Length	Width	l × w	Number of cm²
A	2	3	6	6
B	6	8	48	48
C	5	3	15	15
•	•	•	•	•
•	•	•	•	•
•	•	•	•	•
G	4	8	32	32

Tell the children to inspect the table for patterns. Ask: "Can you find the number of cm² in a rectangle without using the cm² grid paper?"

Discrete Standard Measurement of Volume and Capacity

Each activity that follows involves children in determining the volume of various containers using standard units of volume.

1. Provide the children with a number of cubes, 1 cm on each edge, and a centimeter ruler. Ask the children to find the length, width, and height of the cubes. Inform the children that a cube this size is called one cubic centimeter or one milliliter. During subsequent activities, use the terms interchangeably, so that the children will be able to use either one with facility.

2. Provide the children with a number of small boxes and an abundance of centimeter cubes. Tell the children to carefully stack the cubes into the boxes in order to determine how many will fit into each box. They should record their results in tabular form, using both cubic centimeters and milliliters as their labels.

Tell the children to make several "boxes" of their own by stacking the centimeter cubes. Suggest that no edge should be more than 5 cm long. With a centimeter ruler, measure the length, width, and height of the "boxes." Also count the number of cubes in each "box." Record the results in a table.

Box	Length	Width	Height	$l \times w \times h$	Total Cubes
A	2	3	3	18	18
B	2	2	3	12	12
C	2	4	2	16	16
•	•	•	•	•	•
•	•	•	•	•	•
•	•	•	•	•	•
G	5	2	2	20	20

Ask: "Are there any patterns in the table? Can you find how many cubes there are in a box without counting them?"

4. Provide the children with 50 ml graduated cylinders and about 20 cubic centimeter units. Tell the children to put exactly 20 ml of water into the cylinder.

Ask the children to submerge five cubic centimeters in the cylinder and to record the new water level in milliliters. Add five more cubic centimeters and again record the level. Ask the children to hypothesize about their results. Encourage additional experimentation to verify that one cubic centimeter displaces exactly one milliliter of water in the graduated cylinder.

Photo 6–15

Discrete Standard Measurement of Mass

Each activity that follows involves children in determining the mass of various objects using standard units of mass.

1. Provide the children with two or three hundred cubic centimeters that have a mass of exactly one gram each. Also provide numerous everyday household objects that are appropriate for using on the balance. Let the children balance each object and count the number of grams required to balance it. They should record their results in a table.

This activity will become tiresome very quickly, since the children may be determining the mass of objects as heavy as 200 or 300 grams, and each gram mass must be counted separately.

2. Provide the children with two or three hundred cubic centimeters that have a mass of exactly one gram and a standard set of masses containing pieces of 50, 20, 10, 5, and 1 gram. Let the children verify that the standard masses are correct by balancing them with the cubic centimeters.

3. Provide a standard set of masses and a number of household objects for which to determine the mass. Demonstrate the most efficient method of using the masses to arrive at a correct total. That is, testing the larger masses on the balance and gradually working down to the one gram masses.

Let the children determine the mass of a large number of objects and record their results in tabular form.

4. Provide two wide-mouth one liter containers, a set of masses equaling one kilogram, a balance, and a supply of more than one liter of water. Place one container on each side of the balance. Use the balance adjustment to assure that the containers balance.

Place one kilogram of mass into one of the containers. Ask the children how much water they think it will take to balance the one kilogram mass.

Slowly pour water into the liter container opposite the one kilogram mass. Stop pouring just as the scale balances. Encourage the children to discuss what they have just observed. Ask if they can predict how much water would balance a 500 gram mass. Let them experiment until they are convinced that one milliliter of water has a mass of exactly one gram.

Discrete Standard Measurement of Time

Each activity that follows involves children in determining time using standard units of time.

1. Bring a small pendulum clock to class. Let the children compare the swings of the pendulum with the class pendulum (activity 3, page 219). How many times does the clock pendulum swing during one swing of the big pendulum? Is it always the same number? Does the clock pendulum always swing the same distance, or does it gradually swing shorter and shorter distances as the big pendulum does? When the big pendulum swings a shorter distance, does it take less time for a swing than when it swings a long distance?

2. Bring a wind-up alarm clock to class. Carefully remove the works from the clock, leaving the hands in place. Let the children watch the rotations of the escapement wheel and compare it to their classroom pendulum and the pendulum clock in activity 1.

Discuss how the hands on the clock are only a record of the number of swings of the escapement wheel.

3. Point out the second hand on the classroom electric clock. Discuss how it counts the number of pulses of electricity from the power station.

4. Bring a quartz (digital) watch to class, and discuss how it counts the vibrations of a piece of quartz that pulses thousands of times per second. The quartz watch is more accurate, since the basic unit of measurement is much smaller than for the other time measuring devices.

Generalize from activities 1 through 4 that all clocks are devices that count and record some regular repetitive event.

Photo 6–16

5. Have the children count from 1 to 12 as the classroom wall clock's second hand passes around the clock. After the children have become proficient, have them count from 1 to 60 as the second hand passes around the clock. Next, point out how far the minute hand has moved during the time the second hand passes around the clock five times, ten times, and so on.

After doing the same for the hour hand, introduce the standard terms used for reading and recording time.

Discrete Standard Measurement of Temperature

Each activity that follows involves children in determining temperature of various materials using standard units of temperature.

1.   Obtain a Celsius thermometer that has a temperature range from at least 0° to 100°. Let the students look closely at the scale on the thermometer, noting that its scale ranges from 0° to 100°. Let the children note the present temperature on the scale and speculate how hot 100° would be and how cold 0° would be.

2.   Provide a one-liter container with water and several ice cubes in it and another one-liter container filled with boiling water. The boiling water can be heated with a small coffee cup heater available for about one dollar in most grocery stores.

While the water is heating, let the students observe and record the temperature of the ice water. Stirring and adding ice cubes may be necessary to assure that the thermometer will read exactly zero degrees.

After the heated water has begun to boil, insert the *prewarmed* thermometer into it, and allow the children to observe the 100° temperature. **Caution!** Do not take the thermometer directly from the ice water to the boiling water, or from the boiling water to the ice water. Even a Pyrex thermometer will shatter under those conditions.

3.   Obtain an inexpensive thermometer and mount it outside the classroom window, so that it can be read from inside. Be certain that the thermometer is never in direct sunlight. Have the children regularly observe outside temperatures and record them in tabular form.

The various activities described in this chapter have been presented in a sequential arrangement corresponding approximately to children's ability to learn measurement concepts. The sequence of activities was graphically depicted in Figure 6–1. The teacher is encouraged to develop measurement activities carefully; the results will mean a long-lasting understanding of all types of measurement.

Extending Yourself

1.   Choose a measurement topic from each of the four levels of measurement in Figure 6–1. Write a lesson plan for each. Each lesson plan should contain several activities appropriate to the stage of development under consideration.

2   Write a two page report on "Readiness for Learning Measurement." Include in your report a discussion of the Piagetian concepts of conservation of length, volume, mass, and area.

3.   Investigate one or two standard textbook series for children. Find the measuring topics, and write a list of the topics in the order that they appear in the books. Contrast your list with the order of topics found in Figure 6–1.

4.   Many parents expect their children to be taught how to tell time in kindergarten. Write a brief rationale for waiting until the latter first grade to teach children how to use a clock. Include a list of prerequisite skills for telling time with a standard clock.

Bibliography

Kidd, Kenneth P., Myers, Shirley S., and Cilley, David M. *The Laboratory Approach to Mathematics.* Chicago, Illinois: Science Research Associates, 1970.

Minnesota Mathematics and Science Teaching Project. Units 5, 10, 12, 16. Minneapolis, Minnesota: University of Minnesota, 1969.

Underhill, Robert G. *Teaching Elementary School Mathematics.* Columbus, Ohio: Charles E. Merrill Publishing Co., 1972.

C H A P T E R

Children Developing Through Problem Situations

Problem-Solving Processes

A review of the current literature on mathematical problem solving reveals that this phrase is used in two distinctly different manners. The most usual manner refers to the solution of verbal or word problems. Thus, a statement such as "Johnny had 7 marbles, but lost 3. How many are left?" is considered to be a mathematical problem-solving activity for children.

The second manner in which the phrase *mathematical problem solving* is used in the literature is far more general and encompasses much more than the solution of word problems. George Polya, in his book *Mathematical Discovery*, defines problem solving as the conscious search for some action appropriate to attain some clearly conceived, but not immediately attainable aim.

This second view of problem solving is far more useful as a guide to preparing children to be effective problem solvers in everyday life. It is necessary to provide children with the ability to seek out and perceive patterns and meaning in their everyday environment and to use that meaning to exercise some degree of control over their environment. If there are certain standard mathematical patterns and relationships directly applicable to an everyday problematic situation, it is desirable that children know the patterns and have the skills necessary for applying them.

For example, consider the following problem presented in typical verbal form: "Karen sold 8 boxes of candy and Geri sold 3. How many boxes must Geri sell in order to sell as many as Karen?" It is not difficult to teach children the mathematical pattern that applies to this problem. Almost any six- or seven-year-old can solve 8 − 3 = _____. Discovering which particular pattern applies to a particular problem, however, is far more difficult for children than simply finding the answer once the pattern is known.

Moreover, children will often confront problematic situations to which no previously memorized mathematical pattern applies. Thus, children must also learn to deal effectively with problem-solving tasks that require the invention of solution techniques.

Benjamin Bloom's *Taxonomy of Educational Objectives for the Cognitive Domain* (1956) describes six basic levels of cognitive educational objectives:

Knowledge objectives are objectives for learning rote facts. At this level of cognition, the only problems that can be solved are problems that are identical to problems already solved; that is, both the problem and the solution are known by rote.

Comprehension objectives are objectives for restating, translating, or changing the form of knowledge already known. At this level of cognition, a problem can be solved if it requires the same solution processes as some problem already known.

Applications objectives are objectives for recognizing when a known procedure applies to an everyday situation and applying it appropriately. At this level of cognition, a problem can be solved by a solution technique already known.

Analysis objectives are objectives for learning how to take a complex communication or relational structure and breaking it down into its elemental parts. At this level of cognition, complex problems can be solved even where no applicable solution technique is known in advance. The problem is analyzed and broken into component parts until the structure of the problem is understood. Then a unique solution technique is invented for the problem.

Synthesis objectives are objectives for learning to take bits of data and organizing them into a meaningful structure. At this level of cognition, complex problems can be solved when no applicable solution technique is known in advance. Information about the problem is collected and organized into a pattern or relational structure until a generalization that explains all available data is formulated.

Evaluation objectives are objectives for learning how to compare established criteria to a structural entity and to determine whether the entity meets the criteria. At this level of cognition, complex problems can be evaluated to determine whether a proposed solution technique is appropriate, and a proposed solution can be evaluated to determine if it is correct.

Children should be taught to solve problems at all six levels of cognition. In the typical classroom, only the first three of the levels are taught. In a research project reported in *The Arithmetic Teacher* (May, 1972), it was found that problems in standard textbooks used in elementary school classrooms in the United States are almost exclusively classified in the lowest three levels of the taxonomy. Only 5 percent of the problems could be said to engender cognitive activity at the analysis, synthesis, or evaluation levels.

If problem solving is to be taught in the most effective possible manner, it is essential that the children be exposed to numerous activities for each of the six levels of cognition. Since most standard textbooks contain copious amounts of problems that engender lower cognitive level activity, this chapter will use as examples problems that may generally be classified as analysis, synthesis, or evaluation.

*The Thought Processes Involved in
Problem Solving*

If children are to engage in problem solving at the higher levels of cognition, what mental processes must they be capable of performing? Much research has been conducted related to the problem-solving abilities of children. Unfortunately, much of the research has been preoccupied with the abilities of children to solve word problems.

Suydam and Weaver (1970), isolated eight factors that seem to characterize high achievers in verbal problem solving. The factors are:

1. Ability to note likenesses.
2. Ability to note differences.
3. Ability to understand analogies.
4. Ability to visualize and interpret quantitative facts and relationships.
5. Understanding of mathematics terms and concepts.
6. Skill in computation.
7. Ability to select procedures and data.
8. Reading comprehension.

The first four factors are clearly processes associated with problem solving and are probably best classified at the higher levels of cognition. Factors five through eight are lower cognitive skills of knowledge, comprehension, and application, and are associated only with rote problem-solving activity.

Schweiger and Wheatley (1975) investigated the basic thought processes involved in problem solving. Their conclusion was that eight basic processes are used in problem-solving activity:

1. Classification.
2. Deduction.
3. Estimation.

4. Pattern Generation.
5. Hypothesizing.
6. Translation.
7. Trial-and-Error.
8. Verification.

Young children are capable of mastering all of these processes with the possible exceptions of deduction and hypothesizing. The children's ability to deduce and hypothesize accurately depends partly on the degree of abstraction of the problem presented to them. If the problem consists of tangible, concrete components that the children can sense directly, they are often able to formulate accurate deductions and hypotheses. If, however, the problem is presented in the abstract mode, children will often deduce and hypothesize inaccurately—almost randomly.

Word problems are poor tools for teaching these eight problem-solving processes to young children. There is usually too little data (often only the exact amount needed) to give the child an opportunity to classify. Because the word problem is abstract, deduction is difficult for most children. Estimation may be used, but the expectation of a single correct answer discourages its use. There is usually too little data given to allow pattern generation. Hypothesizing is difficult because of the level of abstraction. Opportunity for translation is extensive, since the child is expected to translate the word statement into an equation. Trial-and-error is discouraged by the expectation of a single correct answer, which may be any one of an infinite set of numbers.

Verification is encouraged when students are asked to check their answers to word problems.

Thus, word problems may be effective for the development of the translation and verification processes. However, word problems are ineffective tools for the development of the other six problem-solving processes.

Further, children should be encouraged to be creative in solving problems. Torrence (1962) has described four areas of creativity that should be explored:

1. *Fluency*—How many possible solutions can be generated?
2. *Flexibility*—How many different types or categories of responses are there?
3. *Elaboration*—How detailed is the solution?
4. *Originality*—Is the solution unique?

Of course, all solutions should be applicable to the problem, but the problems should allow for one or more areas of creativity.

Developing the Problem-Solving Processes

Young children can learn basic problem-solving processes, if they are given the appropriate experiences. The teacher must be careful, however, to fit the problems to the developmental level of the children by choosing problems needing an appropriate level of abstraction. The manner of presentation is also important. The problem should be simple enough to assure that the children can attack it without becoming discouraged, but difficult enough to assure that their analytic abilities will be challenged.

Thus, a good problem for children:

1. Raises some question easily comprehended (though not necessarily easily answered) by children.
2. Provides them with a means (usually concrete) to gather information about the problem.
3. Gives them just enough direction to prevent them from quitting because of excessive frustration.

To assure that children have maximal opportunity to acquire the basic mathematical problem-solving process, they should be exposed regularly to problematic situations, specifically designed to engender the development of one or more processes. It is not always possible to find a problem that will engender the development of one or more of the eight processes previously listed, but there are certain basic characteristics of problems that tend to do so.

First, the problem should have few, if any, abstract components. That is, the relationships in the problem should be presented in a visable and tangible way. If abstractions in the problem have not been previously fully assimilated by the child, the development of the problem-solving process will be inhibited.

Second, the problem should require several steps to a solution. This causes the children to organize, reflect, and record intermediate answers. It also helps to dispel the common notion that a person who is good at mathematics can simply look at a problem and know immediately how to solve it. The children should have the experience of groping through a problem, taking wrong turns, making corrections, and eventually arriving, by some relatively indirect path, at a correct solution.

Third, the ideal problem has multiple correct answers rather than one single correct answer. This allows the children to develop the desirable attitude of continuing to search for additional information, even after the problem has apparently been solved. Also, children can get more problem-solving experience if they know that there are other answers for which they can search.

Fourth, the problems should require analysis and synthesis of information. Simple one-step problems that can be solved by applying a known algorithm do not prepare children for the complex problem-solving situations that they will face in everyday life.

To illustrate these four characteristics, consider the following problem:

The children are given the network shown and a set of attribute blocks. They are asked to place an attribute block in each region on the network, so

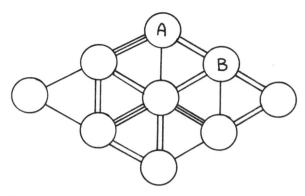

FIGURE 7–1

that each block will be different from the block in an adjacent region by exactly as many attributes as there are lines connecting the regions. For example, if a red, large, triangle were placed in region A, a blue, small, triangle could fit in region B; for it is different from the red, large, triangle in exactly two ways (see Figure 7–1).

As additional blocks are added to the regions, the problem becomes more complex, since some blocks must correctly relate to as many as three other blocks.

This problem has most of the desirable characteristics previously mentioned and engenders most of the eight problem-solving processes previously listed. The relationships in the problem are visible and tangible. The only part of the problem that is somewhat abstract is the necessity of counting up to three lines on the diagram and designating that as the number of attributes needed.

The problem requires several steps for a solution, and the answer is not immediately obvious. There are manifold correct answers to the problem allowing children to find numerous unique solutions. After a few blocks have been placed on the network, each additional region must be carefully analyzed to find the block that meets all of the requirements.

Children's First Experiences with Problems

Clearly, every problem presented to children cannot exhibit every one of the desirable characteristics. However, teachers should be aware of the processes associated with problem solving and should attempt to provide children with numerous experiences that engender each process. The earliest problem-solving experiences involve distinguishing between observations and inferences and observing likeness and difference relationships among objects in the environment.

Photo 7–1

Observing, Inferring, and Comparing

The following are activities that aid in the development of creativity, observation, inference, and comparison skills.

   1.   Give several children a small object, such as a plastic animal. Ask the children to list all the characteristics of the object they can think of. Let the children do this individually first and then compare their answers. Discuss the difference between an observation and an inference. Let the children try this activity with their eyes closed by just feeling the object.

   2.   Let children compare objects drawn out of a bag. Fill a bag with small familiar household objects, and draw one at random. Let a child draw the next object, without looking in the bag, and give one way in which the two objects are alike. The next child to draw an object must tell one way in which

Photo 7–2

his object is like the preceding one. Continue until each child has had a chance. To make this activity more difficult, have the children give a different likeness each time.

To vary the activity, let the children line themselves up by telling one way in which they are like the child in front of them. Older children may do this activity by comparing pictures or numbers.

As children begin to solve problems, their formal experiences should involve classifying objects in the environment, copying structural patterns with objects, and investigating concrete relationships by trial-and-error.

Classifying Objects

Classification activities for developing early problem-solving skills are not essentially different from the activities given in chapter 2. Some examples of classification activities follow.

Photo 7–3

1. Provide children with a tree diagram and a set of appropriate attribute blocks. Ask the children to take each block and move it along the tree to its correct limb.

The children would take the red triangle, move it up the trunk to the first branch, take the left (red) limb, then proceed to the next branch, take the right (triangle) limb, and leave the block at position A on the tree (see Figure 7–2).

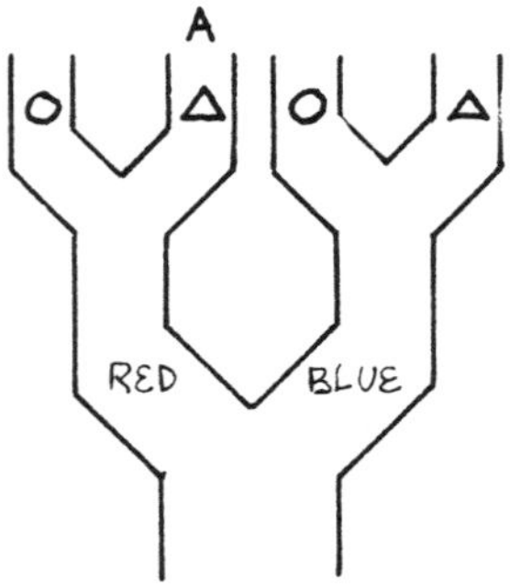

FIGURE 7–2

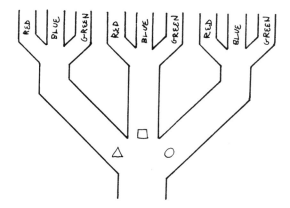

FIGURE 7-3

Other more complex activities may be created by providing a tree that has more options at each branch (see Figure 7-3).

For this tree, a set of nine blocks would be provided. After children have become proficient at working with diagrams of this type, they may be provided with a complex tree and a set of blocks that contain some attributes that are not on the tree. The children must then decide which blocks go on the tree and which do not.

2. Provide the children with a tree diagram that branches into one or two properties and the negation of those properties.

If the children correctly sort the set of large, thick attribute blocks on this tree, there will be only one block in position A, but several blocks on every other limb of the tree (see Figure 7-4).

3. Draw a tree diagram on the floor of the classroom, and label it with attributes of the children, such as boy—girl, tie shoes—buckle shoes, and so on. Have the children walk up the tree one at a time to their correct limb.

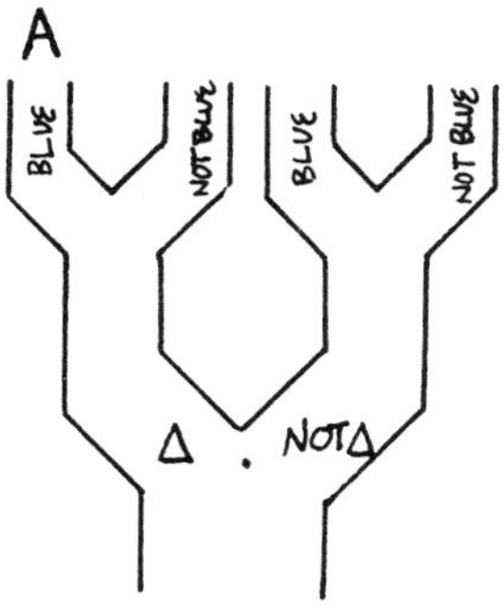

FIGURE 7-4

    4. Draw an unlabeled tree diagram on the board and, with tape, stick attribute blocks on their appropriate limbs. Have the children determine what the label of each limb should be.

Copying Patterns with Objects

The earliest form of pattern activity is simply to indicate that a pattern has been recognized by using similar material and making an exact copy of the pattern. With concrete material, this requires the children to observe the properties of the objects used in the pattern relative to the rest of the objects and to reproduce the pattern by re-creating each of the positional relationships with an identical set of material.

    Thus, the re-creation of a pattern, though it appears to be a simple activity, prepares children to analyze data in much the same way a scientist does when he is attempting to formulate a hypothesis. The relationships discovered by children during these activities will hopefully prepare them for more sophisticated pattern activities, such as filling in missing parts of patterns (interpolation), extending patterns (extrapolation), and inventing patterns (hypothesizing).

    The following activities are examples of making direct copies of a given pattern.

    1. A teacher or child leader should stand in front of the class. Children stand facing the leader and attempt to mimic every move made by the leader. The leader may: touch right shoulder with left hand, make right angle with right arm in front of body, make a "T" in front of body with both hands, bend at the waist extending one foot rearward and one hand forward, and so forth. The activity is limited only by the leader's imagination and always results in some hilarious contortions.

    2. Provide children with a set of wooden or plastic beads having many varied shapes and colors and some shoe strings. Hang some string patterns in front of the children and ask them to make exact copies of them.

    This activity is quite simple, since it involves only a one-dimensional pattern. The relative difficulty of the activity can be increased by using more different shapes, more different colors, and by increasing the period of the pattern; for example, red, blue, green, red, blue, green, . . . has a period of three. Nonperiodic patterns may also be used and are harder than periodic patterns.

    3. Provide children with a set of Pattern Blocks and a pattern already constructed with the Pattern Blocks. Ask the children to make a copy of the pattern by laying their blocks directly on top of the given pattern. When the children are proficient at making a copy on top of a given pattern, have them make an exact copy of the pattern beside the given pattern.

    After the children seem quite confident at making copies of given patterns with pattern blocks, give them a relatively simple pattern and allow one

Photo 7–4

minute for the children to study it. Then remove the given pattern and ask the children to make a copy of it by memory. This requires that the children look at the pattern much more carefully and see it as a whole unit rather than simply look at only one small part at a time.

4. Provide children with a set of Parquetry pieces (see Appendix) and pictures cards. Ask the children to make the Parquetry Patterns by laying the pieces directly on top of the pictures. Later, have the children make copies of the pictures by laying the pieces beside the pictures. Eventually, the children should be able to study the pictures and make a pattern like the picture from memory.

Working with the Parquetry pieces is a slightly higher level of abstraction than working with the pattern blocks as in activity 3. Since the stimulus material is in the form of color photographs of the materials rather than the actual objects, the children must work at the semi-concrete level. As the children become more proficient, they may be challenged by working at the semi-abstract level. For example, provide them with only a line drawing of the pattern to be constructed with the Parquetry pieces. And, to further challenge the children, give them a pattern having only an outline without internal demarcation.

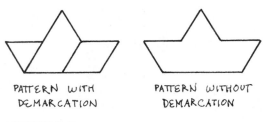

PATTERN WITH   PATTERN WITHOUT
DEMARCATION   DEMARCATION

FIGURE 7-5

Line drawings without internal demarcation are quite difficult to solve and should be given only to children who have demonstrated proficiency at lower levels of abstraction.

5. Provide children with a simple loom and some yarn. Show them several patterns that may be woven on the loom, and let them choose one to produce. Early attempts at copying woven patterns should be limited to variations in only one direction. That is, the vertical lines should be the same color, whereas the horizontal lines may vary in color. Naturally, as the child's proficiency increases, the complexity of patterns attempted should increase.

6. Provide children with a set of attribute blocks and a set of drawings or pictures of one- and two-dimensional arrays of attribute blocks. Ask the children to copy the arrays.

The arrays should start with simple one-dimensional repeating patterns such as that in Figure 7-6.

FIGURE 7-6

Complex two-dimensional patterns such as those in Figure 7-7 should gradually be introduced.

FIGURE 7–7

This activity is a prelude to later activities that are related to the construction of data tables in which the rows and columns in the array are labeled and children put the correct piece in the correct place by reading the labels for the row and column.

As in previous activities, children may begin making copies of attribute block arrays by placing blocks directly on the given drawings. Then, step by step, they will make copies beside the drawings and by memory after studying the given drawings.

Activities 7 and 8 are somewhat simple examples of pattern copying, but they also serve to introduce the child to the concept of translation between *isomorphic* systems. Two systems are isomorphic if they have the same internal set of patterns and relationships. This concept is crucial to problem solving because many problems difficult to solve in their given form are quite easily solved if they are first translated into some other equivalent or isomorphic form.

7. Provide the children with a set of blue attribute blocks. With the red blocks, the teacher makes a pattern involving relationships between shape and size. The children copy the same relationships with the blue blocks.

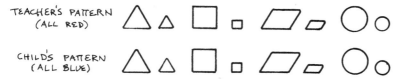

FIGURE 7–8

Next, give the children the large blocks. The teacher makes a pattern with the small blocks and the child copies the pattern with the large blocks.

Numerous other types of attribute materials may be used for similar isomorphism activities. People Pieces and Cuisenaire rods are examples.

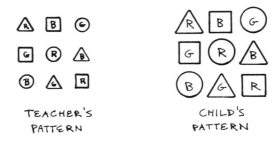

TEACHER'S
PATTERN

CHILD'S
PATTERN

FIGURE 7–9

8.   Provide the children with a set of Pattern Blocks and a mirror that will stand on edge at a right angle to the table or other working surface. Give the children a simple pattern of blocks and stand the mirror along one edge of the pattern. Ask the children to look into the mirror and observe the pattern in the mirror. Tell them to construct the mirror pattern on the back side of the mirror.

MIRROR

FIGURE 7–10

After the children have constructed what they believe to be the correct pattern, they can remove the mirror. If their answer is correct, the constructed pattern will look identical to the pattern in the mirror.

Using Trial-and-Error

Much of children's early problem-solving experiences is trial-and-error because many of the problems young children attempt to solve involve patterns and relationships too complex for them to approach in a systematic fashion. When children learn to crawl and walk, they often attempt to pass through a gap between pieces of furniture that is too narrow for them. Older children might attempt to move the furniture to widen the gap or take a different route to avoid the gap. Younger children will, however, attempt by trial-and-error every possible way to pass through the gap—left side first, right side first, head first, feet first, and usually will end up screaming in frustration at the unsolvable problem.

At any level of development, children and adults may be faced with problems too complex to be solved systematically. Thus, it is important that trial-and-error be encouraged as a legitimate problem-solving technique. The frequent outcome of trial-and-error attempts at problem solving is the discovery of a pattern or relationship that develops into a systematic approach to the problem. This outcome is often the result of what might be called *thoughtful trial-and-error*, which will be discussed later in this chapter.

It is unnecessary for a teacher to contrive activities for children to help them practice trial-and-error. Their lives are full of problems that they will attempt to solve by random guessing. However, examples of experiences children should have include solving picture puzzles, fitting geometric shapes into holes of various shapes, and building towers with odd-shaped blocks.

Just as adult attempts at trial-and-error usually result in the discovery of a systematic approach to a problem, so do the children's attempts result in the discovery of patterns and relationships that will allow them to approach a problem in a systematic way. In general, trial-and-error is a problem-solving approach that enables individuals to gain sufficient experience to avoid solving future problems by trial-and-error.

Later Problem-Solving Processes

After children have had extensive experience with early forms of patterned and structural material, the sophistication of the experiences should be increased. There are several ways to increase the sophistication of the activities. The simplest way is to increase the number of variables children are expected to deal with at one time. Rather than making patterns with two shapes and two colors, children can attempt patterns with three shapes and four colors. Another way is to increase the total number of pieces or components in the patterns that children are studying. Rather than copying a pattern having a total of eight pieces, children can attempt a pattern with twenty pieces. Yet another way to increase the sophistication of the problem is to increase the number of dimensions in the structure of the pattern. Stringing beads is a one-dimensional activity. Parquetry patterns are two dimensional, and building patterns with geoblocks is a three-dimensional activity.

Thus, by controlling the number of variables, the number of components, and the number of dimensions involved in the activities, the level of the activities can be tailored to the experience level of the children.

Classifying Objects with Several Variables

The following classification activities are similar to those mentioned in the previous section of this chapter. However, the number of variables and components has been increased and the activities are more suited for older, more experienced children. The children are also introduced to the Carroll Diagram, which will prepare them for the later skill of organizing data into tables.

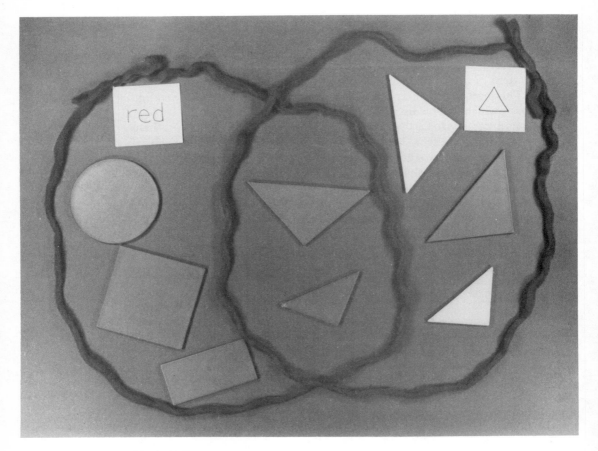

Photo 7–5

1. Provide children with a set of attribute blocks and several large (50 cm diameter) loops of yarn or heavy cord. Ask the children to lay the loops on the floor, and give exercises such as these:

(1) Place the triangles in one loop and the squares in another.

(2) Place the yellow in one loop and blue and green in others.

(3) Place the red in one loop and the triangles in another. (This requires the loops to be overlapped—a condition which the children may not realize at first.)

(4) Lay the loops in the classic Venn diagram pattern, and label them. Ask the children to put the correct blocks in the correct regions in the diagram (see Figure 7–11).

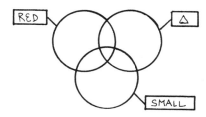

FIGURE 7-11

Let the children work with this type of configuration extensively. Change the labels so the children must rearrange the blocks. Ask which blocks do not fit in the diagram. Place the blocks in the circles appropriately, then remove the labels from the circles. Let the children try to relabel the circles appropriately.

Occasionally, use the negation of one or more of the properties as one of the labels (see Figure 7–12).

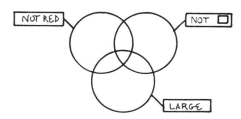

FIGURE 7-12

This gives practice in the use of the complements of sets as well as the sets themselves. Also, if the blocks are placed as in Figure 7–12 and the labels are removed, the correct relabeling of the diagram is a very difficult problem for children to solve.

After children have had numerous experiences with the Venn diagram and can do the activities with proficiency, change the diagrams, so that they begin to look like Carroll Diagrams or data tables. Examples are given in activities 2, 3, and 4.

2. Make a diagram such as the one in Figure 7–13 and provide the children with a set of about sixty buttons. Let the children sort the buttons into the separate regions. After the buttons are sorted, give students a mimeographed sketch of the diagram and have the students record the number of buttons in each region. If the students do not have sufficient number skills for the task, they can record the numbers by making tally marks in the regions.

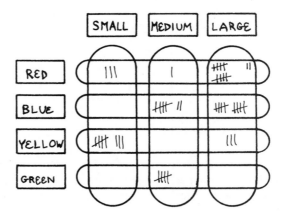

FIGURE 7–13

If the students have sufficient number skills, the teacher can have them refer to their diagram (table) and answer questions such as: "How many blue large buttons are there?" "Are there more small, red or green, medium buttons?"

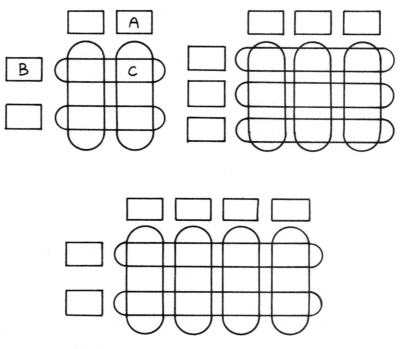

FIGURE 7–14

3. Make up several posterboard diagrams such as those in Figure 7–14. Make label cards for each of the properties of the attribute blocks and their negations. Provide groups of the children with one or more diagrams, a set of attribute blocks, and a set of label cards. Let one child in the group place the labels and the rest of the group put the blocks in the appropriate regions.

The children will learn by experience that two different values of the same variable must not be aligned across or down. For example, if "square" (a value of the variable shape) is placed at point A in Figure 7–14, then "triangle" (also a shape) cannot be placed at point B. If the labels are placed in this manner, then the blocks placed in region C in the diagram would be both triangles and squares. Clearly, no such block exists. The teacher should oversee the activity at first to assure that students do not make mistakes of putting *either* triangles *or* squares at C. This error could lead to confusion when the students later attempt to table data. If problems of understanding arise, use activities such as 1, 4 and 5, until the children clearly acquire the concept of set intersection.

4. Make a chart such as the one in Figure 7–15 and provide children with an assortment of household junk appropriate for the chart. Let the children sort the junk into the regions on the chart.

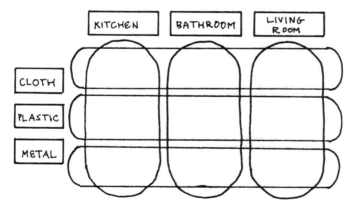

FIGURE 7–15

5. Provide the children with a set of People Pieces and a set of label cards (fat, thin, male, female, red, blue, adult, child). Let them make 2 × 2 arrays and place the labels in their appropriate places.

Let the children make as many arrays as they can find.

Next, provide the children with a chart such in Figure 7–16b. Lay the pieces on the chart in a consistent array, and let the children put the label

FIGURE 7–16a

cards in their appropriate places. Rearrange the People Pieces, and let the children relabel the chart correctly.

After the children can correctly label the chart, place the labels on the chart first and allow the children to place the People Pieces appropriately. As the children become more proficient, let them both label the chart and arrange the pieces in as many ways as they can find.

		Male		Female	
		Fat	Thin	Fat	Thin
Child	Red				
	Blue				
Adult	Red				
	Blue				

FIGURE 7–16b

The children should be proficient enough with the chart to be able to describe exactly the properties of the piece that would fit in a designated square when all of the labels are in place.

*Extending and Filling in Patterns
(Extrapolation and Interpolation)*

Mathematics has been described as a unified set of patterns and relationships among abstractions. However accurate the definition may be, mathematics certainly does contain many patterns. Some patterns must simply be committed to memory. There are far too many patterns, however, to justify trying to commit all of them to memory. Therefore, it is necessary that children learn to in-

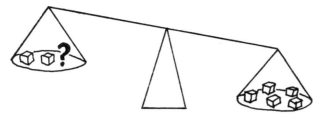

FIGURE 7–17

vent patterns and to complete or extend patterns when only a part of a pattern is given.

For example, $2 + 3 = 5$ is a pattern that children should eventually be expected to commit to memory. However, young children may not have this pattern committed to memory when they first see it in this form: $2 + \underline{\quad} = 5$. Part of the pattern is missing, and children are expected to be able to supply the missing part. If the problem is presented abstractly, as above, children may be unable to supply the missing parts of the pattern, because they do not fully understand all of the relations symbolized.

If the pattern $2 + \underline{\quad} = 5$ is presented in concrete form, however, the children may find the problem quite simple to solve, and, as a bonus, learn the fact by memory (see Figure 7–17).

The ability to perceive a pattern and to supply missing parts is a skill basic to the learning of mathematics. Completing a pattern requires the same skills as solving equations. The equation presents a pattern to the child; there is something missing in the pattern, and the child must fill in the missing parts.

The following activities provide practice in both extrapolation and interpolation. *Extrapolation* involves the understanding of a pattern so that the pattern can be extended beyond the information given. For example, 1, 4, 9, 16, . . . , is a pattern that, when recognized, can easily be extrapolated to include 25, 36 or any of the other squares of the natural numbers. *Interpolation* would be the process of filling in the missing parts of a pattern. In the same example (1, 4, , 16, 25, . . .), filling in the missing 9 would be an example of interpolation.

1. Make up pattern strips from posterboard having approximately ten 5×5 cm squares. Draw or glue objects onto seven or eight of the squares, so that a pattern is formed. The children are asked to fill in or extend the pattern (depending on which squares have been left blank). For example, in Figure 7–18, the pattern on the first strip is "square, triangle, circle, square, triangle, circle," and so on. The second strip is "red triangle, red circle, red square, red diamond," then "blue triangle, blue circle," and so on. The third strip is "bathroom item, kitchen item, bathroom item, kitchen item," and so on. The fourth strip is "wood item, plastic item, metal item, wood item, plastic item, metal item," and so on. The fifth strip is left for the reader to solve.

There are many possibilities for patterns such as these. Children should have as many patterns of this type as the teacher can provide.

2.   Provide children with a set of Cuisenaire rods and a pattern of rods in a stairstep with one or two rods missing. Ask the children to fill in the missing rod. Let the children play in pairs. One of the children turns away while the other child removes one or more rods from the pattern and closes the gap. The other child attempts to identify and replace the missing rods.

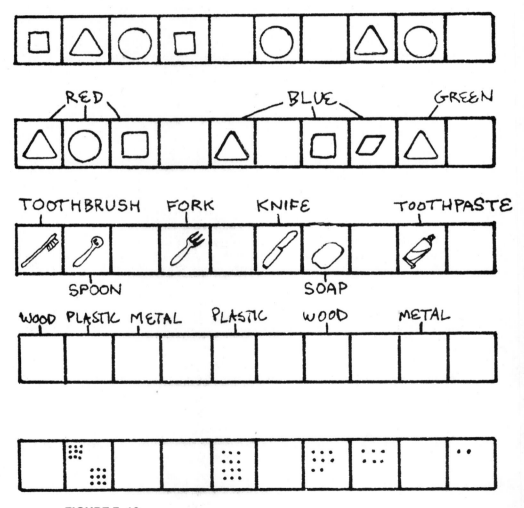

FIGURE 7–18

Photo 7–6

Later, it may be desirable to invent other patterns with the rods to let the children fill in or extend. There are numerous patterns possible.

3. Provide the children with attribute block arrays such as those in Figure 7–19 and ask: "What is missing?" or "What is out of place?" or "What doesn't belong?"

As in previous examples, there are various other possible patterns for the children to solve. Children should become proficient enough with patterns of

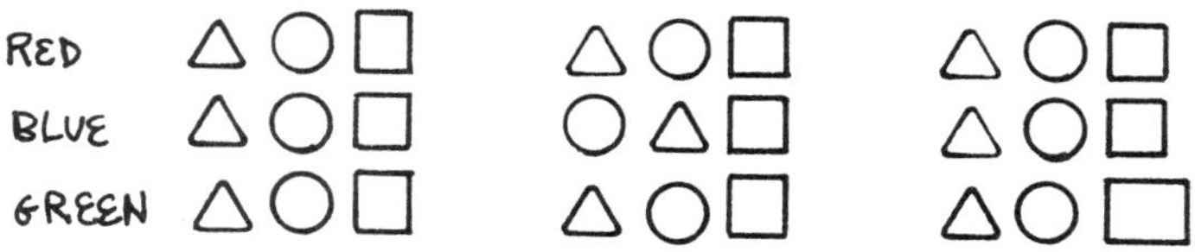

RED   △ ○ □   △ ○ □   △ ○ □
BLUE   △ ○ □   ○ △ □   △ ○ □
GREEN   △ ○ □   △ ○ □   △ ○ □

FIGURE 7–19

this type to be able to invent problems for themselves. Problems of this type can be done with any structured set of attribute material, People Pieces, color cubes, or teacher-made materials.

4. Provide children with a set of attribute blocks and a four-by-four array, such as the one in Figure 7–20 with a few blocks in place. Ask the children to put the remaining pieces in place. Let the children make up different patterns of their own.

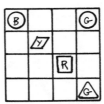 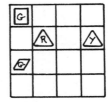

FIGURE 7–20

After the children are proficient at completing such patterns, let them attempt a *Latin Square* in which no row or column may contain two shapes or colors alike. Since this is a very difficult activity, it might be well to begin by making only the rows containing no two shapes or colors alike. Then add columns to the task, and eventually the diagonals.

5. Provide children with a set of cards with images of the attribute blocks or other attribute pieces drawn on them. Arrange the cards in an array such as the ones in activity 4 but with the images of the attribute pieces turned face down. Give the children a set of the actual attribute pieces scrambled in a container.

Let one of the children draw an attribute piece at random from the container. By trial-and-error, the child turns up a card from the array and checks to see if it matches the attribute piece. If there is a match, the card is left visible, and the attribute piece is placed on top of it. If there is not a match, the card is turned down, and the attribute piece is placed back into the container. The object of the activity is to match each card with its appropriate piece.

The children will soon learn to note carefully the properties of the image on the card. Since the cards are in a specific array, each card will tell the children much about where the other cards will be located. As the cards are turned up, they will gradually form a pattern that will allow the children to make the final few choices without error.

Although this activity is intended to give the children practice in observing and completing a pattern, it also is a good activity for teaching another important problem-solving technique—thoughtful trial-and-error.

Thoughtful Trial-and-Error

Trial-and-error is one of the most used problem-solving activities, yet it is rarely a part of the children's regular curriculum. Any experience with trial-and-error usually comes incidentally as an outgrowth of problems that occur naturally in the children's environment.

The key to this section is the word *thoughtful*. It is not sufficient to provide children with experiences in which they make random tries to find an answer. Thoughtful trial-and-error uses each trial to note the size of the error and to attempt to improve the next trial by reducing the size of the error. Subsequent trials are used to further reduce the amount of error until children are satisfied that a correct solution has been reached.

The following activities all involve trial-and-error. Some may result in single correct answers, so that there is no doubt the problem is completed. Others have approximate answers that may be refined until the children are satisfied the answer is close enough.

1. Bring a metronome to class (borrowed from the music teacher) and set it to beat about 50 times per minute. Provide the children with string and weights for making pendulums. Encourage the children to make a pendulum that will beat precisely in unison with the metronome.

Suggest that the children attach their pendulum to a table or other fixed object. Swinging it from their hands will cause the rate to vary. Much experimentation will be necessary to vary the length and get the pendulum to beat in perfect rhythm with the metronome.

2. Provide the children with a set of eleven red pieces of posterboard 2 cm wide and varying in length from 2 through 12 cm. Provide them with an identical blue set. Ask the children to put a red piece and a blue piece together to make a red-blue train. Have the children make other red-blue trains the same length. Finally, tell the children to try to make red-blue trains all the same length using every red and blue piece they have.

Much trial-and-error experimentation will ensue until the children realize that all trains in the final solution will be 14 cm long. The red 2 cm piece must be matched with the blue 12 cm piece, the red 3 cm with the blue 11 cm, and so forth.

3. Make a sliding pan balance with a meter stick and hooks made from coat hanger or other stiff wire (see Figure 7–21). Use styrofoam cups or small aluminum foil pans for the balance pans. Provide children with a set of uniform masses of approximately 15 grams each.

FIGURE 7–21

Place one mass on the right pan and two masses on the left pan. Ask the children where the pans must be placed to achieve balance. They may solve the problem by trial-and-error, but if they have adequate number skills they should record the marks where the pans balance.

Try the activity with three masses on the left pan, one on the left and two on the right, two on the left and three on the right, and so forth. Let the children make up new problems of their own.

4.   Using a fixed pan balance, give children six wads of clay having masses of 10, 15, 20, 25, 30, and 35 grams. Ask the children to divide the wads among three people, so that each of the three people will have the same amount by mass. The children will experiment with different wads on the balance, until they have paired the 10 and 35, 15 and 30, and 20 and 25 gram masses.

5.   Find a relatively tall, narrow jar that has a mouth large enough for the children to reach into. Select a set of seven or eight smooth stones that vary in volume between 30 and 500 cubic cm and will fit into the jar. Put enough water into the jar to cover the largest stone, if it were lying on the bottom. Mark the water level "start" for future reference.

Next, put two or three stones in the jar, mark the new water level, and label the mark A. Put another combination of stones in, and mark level B. In the same way, mark levels C and D. Make certain the levels are at least two centimeters apart.

Provide the children with the jar filled to "start" and the stones. Tell the children to find the combination of stones that will make the water rise to each mark. Tell them to be certain the water is at the starting level before they begin each experiment.

If the children are able to write, label the stones and have them record each trial to avoid repetition of trials. After the first few trials, the children should begin to select their stones carefully according to how much volume is needed to achieve the desired water level.

Translating Between Systems
(Isomorphisms)

Much of mathematical problem solving is done by finding similarities between a known mathematical system and an everyday problem. Once the similarity is found, the everyday problem can be translated into the mathematical system and solved mathematically. Thus, one very important problem-solving process is the ability to translate between systems.

Of particular interest in mathematics are similar systems that are said to be *isomorphic.* Two systems are isomorphic if: (1) there is a one-to-one relationship between the systems, and (2) the relationships among the members of the first system are the same as the relationships among the members of the second system. For example, consider the isomorphism between the Brown family and the Adams family.

Mr. Brown—Mr. Adams
Mrs. Brown—Mrs. Adams
Jack Brown—John Adams
Susan Brown—Cheryl Adams

There is a one-to-one relationship between the Brown and the Adams families, as can be seen from the illustration. Moreover, the relationships in the Brown family are exactly identical to the relationships in the Adams family. For example, Mr. Brown is the husband of Mrs. Brown, and Mr. Adams is the husband of Mrs. Adams. Susan Brown is the sister of Jack Brown, and Cheryl Adams is the sister of John Adams.

If the Brown family had two boys in it, rather than a boy and a girl, then there would no longer be an isomorphism between them, because the relationship between the two Brown brothers would not be the same as the Adams brother-and-sister relationship. Moreover, if one of the families had a fifth member, there would not be an isomorphism between them, because the families would not have a one-to-one relationship between them.

The following activities each establish at least two mathematical systems having an isomorphism between them. The children are asked to observe a relationship in one of the two systems and to translate the relationship into another system. Hopefully, as the children become proficient at translation and more familiar with standard arithmetic systems, they will translate everyday problems into isomorphic mathematical systems.

1. Let the children work in pairs. Give one of the children the set of large attribute blocks and the other the set of small blocks. The first child lays out a pattern of his own invention with the large blocks. The second child attempts to make exactly the same pattern with the small blocks.

The children may also wish to play the game with the red and blue blocks. One child makes a pattern with the red blocks, and another copies it with the blue blocks.

2. Let the children work in pairs. Provide them with a set of Cuisenaire rods. The children make a stairstep pattern with the Cuisenaire rods. Then one child faces the pattern and puts his hands behind his back. The other child places a rod in the first child's hand. The child holding the rod attempts to identify the color of the rod by feeling its length and comparing it to the stairstep pattern.

3. Provide the children with a graphic display of the Brown and Adams families previously described. Ask questions such as: "Jack is related to Susan as John is related to whom?" "Mrs. Adams is related to Cheryl as Mrs. Brown is related to whom?"

4.   Mount an array of attribute blocks onto a 30 × 30 cm piece of cardboard, and place it in a large opaque bag. Give the children a set of attribute blocks, and ask them to make an array like the one in the bag by feeling the array through the bag.

5.   Mount a set of 10 or 12 different color squares on a bulletin board in a random pattern. Make a set of 10 or 12 circles so that there is one circle the same color as each square. Mount them randomly among the squares on the board. Attach a piece of yarn with a paperclip on the end to each square.

Ask the children to find the circle that goes with each square and clip the yarn to it. The same activity may be done using shape as variable rather than color.

6.   Let the children work in pairs. Provide one of the children with all of the *adult* People Pieces and the other child with the *child* People Pieces. The first child removes an adult piece from the set, and the second child tries to remove the matching piece from the child set.

After a few tries, the children may wish to divide the pieces on the basis of sex, color, or girth rather than age to play the same game.

7.   Make a set of attribute cards such as those in Figure 7–22. Use any appropriate symbols on the cards.

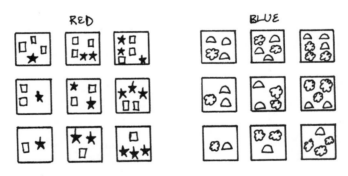

FIGURE 7–22

Give the shuffled red and blue cards to the children and ask them to try to find the red card that matches each of the blue cards.

Lay the two sets of cards in the arrays shown in Figure 7–22, but turn the red cards face down. Point to the back of a red card and let the children explain what is on the card by looking at the matching blue card.

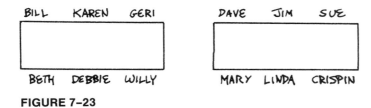

FIGURE 7-23

8. Seat two groups of children around two tables in the room as in Figure 7–23.

Ask relational questions such as: "Bill is related to Karen, as Dave is related to whom?" (Jim). "Debbie is related to Geri, as Linda is related to whom?" (Sue).

9. Make a coordinate system for your classroom using colors and shapes as the variables (see Figure 7–24).

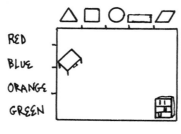

FIGURE 7-24

Along one wall, place six or seven shapes about one meter apart. Along an adjacent wall, place four or five color indicators about one meter apart.

Encourage the children to describe locations in the classroom by using shapes and colors. For example, "The teacher's desk is at the blue triangle." "The bookcase is at the green diamond." When giving classroom directions use the coordinate systems to tell the children where to go in the classroom.

After the children have become accustomed to the system, make a scale drawing of the classroom on a large square piece of posterboard. Affix the colors and shapes in their appropriate places on the scale drawing. Make small posterboard models of the furniture and other objects in the classroom. Let the children attempt to place the furniture in the appropriate places on the scale drawing. Let the children experiment with rearranging the room using the scale model.

Children Applying Problem-Solving Skills

Most of the activities in the previous two sections of this chapter were contrived to create situations in which the children would practice some specific problem-solving skill. The problems were *clean* in the sense that the variables in the problems were carefully controlled. Because of the contrived nature of the material used, there was little opportunity for any real-world variables to contaminate the activities and cause spurious results.

This approach to teaching problem solving may be effective for teaching specific skills, but it is useless unless children are eventually thrust into the real world of approximate measurement, unwanted variables, and unpredictable results. This section of the chapter will assume that the children have developed some or all of the skills introduced in previous sections. It also will assume that the children have developed basic number skills, including addition, subtraction, and simple multiplication.

Classifying Data

Up until this point, the experiences in classification have involved qualitative data, such as color and shape. Most real-world problems will involve the classification of quantitative data. The same processes of classification that are used to organize and systematize qualitative data can be used with quantitative data.

1.   Make a chart such as the one in Figure 7–25 and attach the appropriate labels. Provide a wall chart for measuring height in centimeters. After the children have been measured, let them place a token in the appropriate region on the chart.

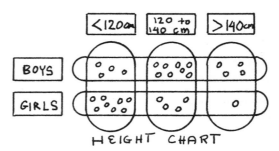

FIGURE 7–25

Ask questions about the chart such as: "Are there more tall boys or tall girls?" "How many boys are there in the class?" "How many girls are there that are more than 140 cm tall?" "How many students are between 120 and 140 cm tall?"

The activity can be varied many other ways by changing the labels on the chart to other descriptions such as "Wears glasses," "Length of shoe," "Length of hair," and so forth. Before the labels are changed, the students may wish to make a permanent record of their data by sketching the diagram, labeling it, and writing the correct number in each space for the number of tokens.

2. Provide children with three or four containers of various sizes, some standard size styrofoam cups, and a chart such as in Figure 7–26. Provide some measuring medium, such as sand, water, or rice. Let the children measure the number of cups that each container will hold, and place the appropriate number of cups on the chart.

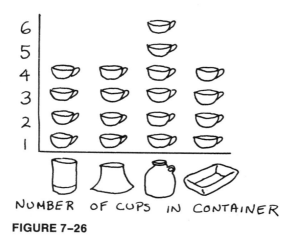

NUMBER OF CUPS IN CONTAINER

FIGURE 7–26

Ask questions such as: "Which container holds the most cups?" "Which holds the least?" "Which container holds two more than container D?"

3. Make a pegboard chart such as the one in Figure 7–27 and provide children with a supply of golf tees. Divide the children into groups of three, and give each group a juice can and a supply of uniform-sized objects such as centimeter cubes. Tell each group to find how many of the centimeter cubes will fit into the can.

After each group has arrived at their number, they take a golf tee and insert it in the pegboard above the number closest to their answer. The answers will vary because of different packing techniques. The golf tees should form a rough bell-shaped curve.

Discuss with the students why the answers were different, and which answers were obtained most often. Do the same activity with other containers. If

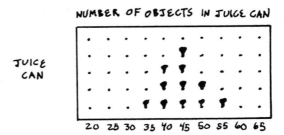

NUMBER OF OBJECTS IN JUICE CAN

JUICE CAN

20 25 30 35 40 45 50 55 60 65

FIGURE 7–27

centimeter cubes are used, rectangular containers will produce answers much nearer the same, since the students will tend to pack the containers the same way.

4. Make a chart such as the one in Figure 7–28 and have the children keep track of how many students are absent on each of the five days of the week.

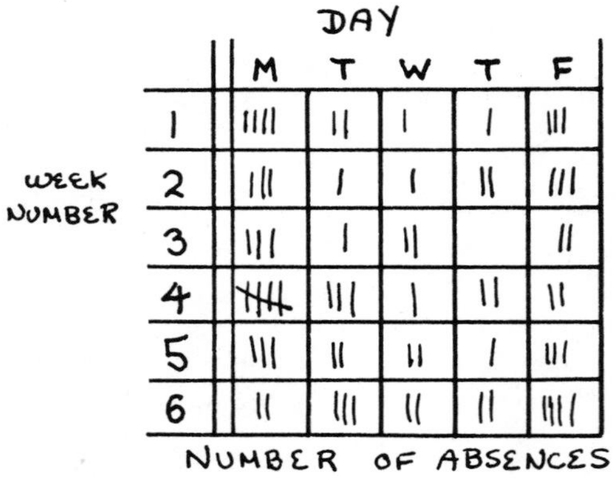

DAY

WEEK NUMBER	M	T	W	T	F
1	IIII	II	I	I	III
2	III	I	I	II	III
3	III	I	II		II
4	THL	III	I	II	II
5	III	II	II	I	III
6	II	III	II	II	IIII

NUMBER OF ABSENCES

FIGURE 7–28

After the chart has been kept for several weeks, ask questions such as: "On which day of the week are people most often absent?" "Least often?" "Have absences increased or decreased through the year?" "What day of what week had no absences?"

Use charts to have students record other daily events, such as weather. The children should become aware that almost any daily circumstance can be charted and organized into tables to allow access to information.

5. Provide children with a small plastic container and a supply of pennies. Provide or have the children make a chart such as the one in Figure 7–29. Have the children put one penny in the container, shake it, and dump it out. If the penny is tails, repeat the procedure. If it is heads, add a penny and shake again. Thereafter, for each penny that comes out heads, add another penny to the container.

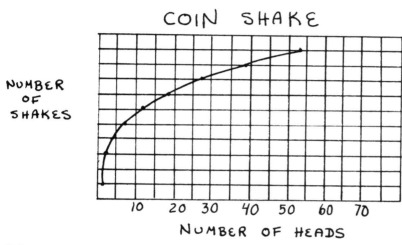

FIGURE 7–29

After each shake, for that numbered shake record the number of heads thrown on the chart. (If a shake yields all tails, shake again, not counting that shake.) After six or seven shakes, ask the students if they can predict how many heads may be next. Ask the students why the graph makes a curving line.

Using Patterns for Estimation

The final activity in the preceding section illustrates how a graphic representation of data may be used to estimate other data not immediately available. Once a pattern has been established, it is possible to make an "educated guess" at additional data by using the processes of interpolation and extrapolation.

In each of the following activities, data are taken from some common everyday occurrences and plotted on a graph. Students are to attempt to draw conclusions on the basis of the data they have graphed. Many unwanted variables will creep into the projects to cause the data to be erratic. The sun will refuse to shine on days when shadows are to be measured. Plant growth will be stunted by too small a container or by too much or too little water. Each of these "contaminating" variables will help to make the problems realistic and will offer the teacher an opportunity to question students regarding the failure of the data to fall into a perfect pattern.

1. Divide the students into groups, and have them plant mung bean seeds. Each day at the same time of day the students should measure the height of the mung bean plants, and record the measurement on their graph (see Figure 7–30). Ideally, several different measurements should be taken and an average measurement recorded on the graph.

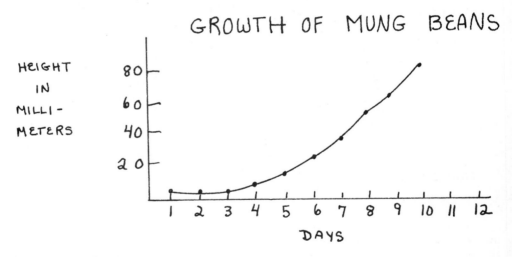

FIGURE 7–30

As a pattern is established, the children should attempt to predict how tall the plant will be the next day or on a Monday after a weekend. They can pencil their predictions in on the graph in one color and record actual growth with another color to make a comparison of the real and predicted growth. Encourage the children to make predictions for a day in advance and a week in advance, and then compare the accuracy of these two predictions.

2. Place an outdoor thermometer outside a classroom window so that the temperature can be recorded conveniently. Make a graph, such as the one in Figure 7–31, and assign pairs of students to make temperature readings every hour on the hour. After the second or third day of recording temper-

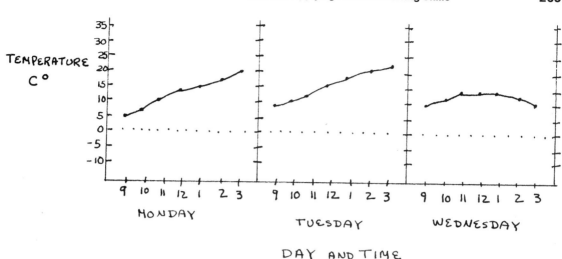

FIGURE 7–31

atures, let one pair of students inspect the graph and attempt to predict the temperature reading, while another pair of students takes the actual readings.

The students will find that it is very difficult to predict the 9:00 A.M. temperature accurately; but once that temperature is known, the other temperatures for the day can be quite accurately predicted. It may be desirable to plot each daily graph on a separate sheet of thin paper, so that they can be overlayed for close comparison of daily patterns. When a day such as Wednesday (Figure 7–31) occurs, ask the the children to watch the television weather report that evening to try to determine why it happened. Try to classify the separate days according to the shape of the graph; for example, sunny, cloudy, rainy, cold front, and so on.

3.   Choose a time of day that will be convenient for several weeks and is also a time when the sun shines through the classroom window. Place a 5 cm piece of masking tape on the window so that its shadow falls on the window sill. Place a 30 cm piece of tape on the window sill, so that the shadow of the 5 cm tape on the window falls in the center of the 30 cm piece on the sill.

Each day thereafter, record the exact time of day that the shadow of the 5 cm piece of tape lies exactly on the 30 cm piece. Plot the time of day on the graph as shown in Figure 7–32.

As in other activities, have the children predict the time of alignment for several days in advance. Careful observation and recording will allow the children to predict the time of alignment to within one minute for several days in advance.

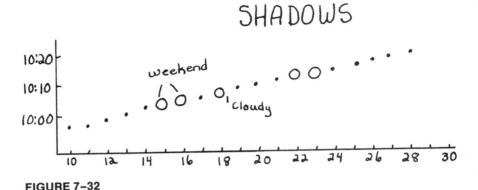

FIGURE 7-32

4. Provide the children with inexpensive ribbon about one centimeter wide, scissors, and an abundance of round objects from 10 cm circumference to 50 cm. On the chalkboard or a piece of posterboard make a large chart at least 60 cm tall and 40 cm wide.

Let the children measure the circumference and diameter of several objects and cut the ribbons in appropriate lengths. Have the children tape each pair of ribbons to the chart so circumference ribbon C extends upward from the circumference index on the chart, the diameter ribbon D extends horizontally from the diameter index on the chart, and the ends of the two ribbons meet at a point P, as shown in Figure 7-33a. After several measurements, the chart will appear as in Figure 7-33b.

The children should note that the points at which the diameter and circumference ribbons meet form a straight line. The line could be drawn in for future reference.

Next take an object such as a basketball that has a diameter that is not directly measurable. Ask the children if they can use their chart to estimate the diameter of the basketball. Encourage them to measure the circumference of

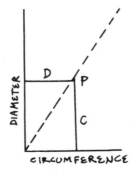

FIGURE 7-33a

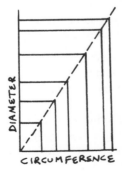

FIGURE 7-33b

the basketball and place the circumference ribbon in the appropriate spot on the chart. From this information, they should be able to cut a ribbon that would fit in the "diameter" spot on the chart.

Using Strategy

Young children are capable of using some fairly sophisticated strategies to solve problems if the problem is presented to them concretely. Developing a strategy involves several of the skills described earlier: observation, classification, deduction, pattern generalization, hypothesizing, translation, thoughtful trial-and-error, and verification.

The following are games that require the children to develop a strategy in order to be successful.

1. Place ten objects in front of two players. Each player may take one or two objects at his turn. Players alternate. The person who takes the last object wins.

Ask the children how they can be sure of winning. Would it be better to be first or second?

Try the game with more or fewer objects. Ask the children if they see a pattern? What are the "key" numbers?

Try allowing players to take up to 3, 4, or 5 objects at a turn.

Photo 7–7

Change the rules so that the person who takes the last object loses.

2. Draw as many patterns as you can using five congruent squares (pentominoes). Each square must share a side with at least one other square. An examples is given in Figure 7–34.

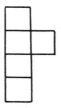

FIGURE 7–34

3. Place two red disks and two yellow disks on the playing strip as shown in Figure 7–35. The object is for one child to exchange positions of the red and yellow disks so the red disks are at the left and yellow at the right (see Figure 7–35). The child may jump one disk over another disk of a different color or slide to an adjacent blank space.

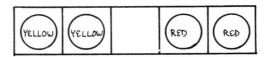

FIGURE 7–35

Let children try this activity with three disks of each color on a longer strip having seven regions. Then let them try it with four or five disks of each color. Ask the children to search for a pattern of moves.

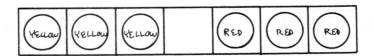

FIGURE 7–35a

4. This is a game called David's Magic Square. Using three sets of the numerals 1, 2, 3, have the children put the numerals in the frame, so that each row, each column, and each diagonal has a sum of six. The picture shows a possible solution because of the diagonal of 2's.

Photo 7-8

5. The game of Pawns, suggested by Kohl (1974), begins with a 3 × 3 playing board and two sets of different pieces or disks at each end (see Figure 7-36).

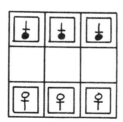

FIGURE 7-36

The object of the game is for two players to move consecutively until one player either: (a) has a piece in the opponent's beginning row, or (b) makes the last move. He is then the winner.

Each piece may be moved one square forward or one square diagonally if he makes a capture. A capture occurs when a diagonal move puts the player's piece in a square occupied by the opponent. No capture can take place on a forward move. The opponent's captured piece is removed from the board (see Figure 7–37).

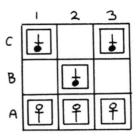

FIGURE 7–37

The first player, whose pieces are in row C, has moved a piece from C-2 to B-2.

The second player, whose pieces are in row A, may either move to position B-1, B-3, or he may move diagonally to B-2 and capture the piece that is already there.

Once basic strategies are established, players may wish to use the same rules with a board that is 4 × 4, or 5 × 5, or even larger.

Conclusion

The famous French mathematician and philosopher, René Descartes, proposed a universal method for solving any problem that mankind might ever face. His four simple steps were (Polya, 1962, p. 22):

1. Reduce the problem to a mathematical one.

2. Express the mathematical problem algebraically.

3. Write it as an equation.

4. Solve the equation.

Even though we seem to be able to teach the last two steps of this universal technique quite effectively, much difficulty is encountered with the first two. It is virtually, if not completely, impossible to reduce some problems to mathematical terms. We can, however, learn much about a problem by following a few basic steps.

1. If possible, first look at some simple form of the problem. If a simple form of the problem can be understood and solved, perhaps enough will be learned to allow a more difficult form to be solved.

2. Try to gather as much information about the problem as possible. Use trial-and-error, if necessary, to gather data.

3. After data have been collected, organize or classify the data so that structural aspects of the problem will become more apparent.

4. Search for patterns in the data. Table, graph, or diagram the data to reveal the patterns that are present.

5. Use the patterns discovered to predict additional data and to fill in gaps or extend the patterns.

6. Verify the predicted data by checking to see if it works in the problematic situation.

7. Try to use the information learned in the above steps to hypothesize a generalization about the data that will solve all problems of the same type as the one just solved.

8. Verify the hypothesis.

Children should have many experiences with problems that will cause them to engage in these eight steps and the processes they imply. The problems in this chapter have been written with these steps in mind. There are, of course, far too few listed here to fill the needs of an early childhood curriculum; but everyday life is an infinite resource.

Extending Yourself

1. Check through several mathematics education methods books and observe the point of view taken by the book on problem solving. Try to determine whether the point of view best fits the first or second one expressed in the opening paragraphs of this chapter.

2. Reread the brief discussion of the "Taxonomy of Educational Objectives" early in this chapter and, if possible, read the description of the Cognitive Domain from the original source. Then, from any standard textbook for first or second grade, randomly choose ten pages from the book and classify the activities on the ten pages at the different levels of the Taxonomy. Determine the approximate percentage of activities for each level.

3. Study the characteristics of good problems listed and discussed in this chapter. Use children's texts, mathematics methods books, and puzzle and game books to find ten problems for children that exemplify one or more of the characteristics listed.

4. Make a set of materials for the problem #3 described on p. 268. Try to work the problem using the minimum number of necessary moves. Make a table and record the number of slides and jumps required for each solution. Try to predict how many moves would be required if there were 20 disks on the board.

Bibliography

Bloom, Benjamin S. (ed.). *Taxonomy of Educational Objectives: Cognitive Domain.* David McKay Company, Inc.: New York, 1956.

Callahan, LeRoy G., and Passi, Sneh Lata. "Textbooks, Transitions, and Transplants," *The Arithmetic Teacher,* Vol. 19, No. 5 (May, 1972), pp. 381–385.

Kohl, Herbert R. *Math, Writing, and Games.* New York: The New York Book Review, 1974.

Polya, George. *How to Solve It.* Princeton, New Jersey: Princeton University Press, 1945.

———————————. *Mathematical Discovery.* New York: John Wiley and Sons, Inc., 1962.

Riedesel, C. Alan. *Guiding Discovery in Elementary School Mathematics.* New York: Appleton-Century, Crofts, 1967.

Schweiger, Ruben D., and Wheatley, Grayson H. "Basic Thought Processes in Mathematical Problem Solving." Paper presented at the national meeting of the National Council of Teachers of Mathematics, Denver, Colorado, April, 1975.

Suydam, Marilyn B., and Weaver, J. Fred. "Verbal Problem Solving," *Interpretive Study of Research and Development in Elementary School Mathematics.* (Grant # OEG–O–9–480586–1352 (010)), The Pennsylvania State University.

Torrence, E. Paul. *Guiding Creative Talent.* Englewood Cliffs, New Jersey: Prentice-Hall, Inc., 1962.

C H A P T E R

Preparing Children's Learning Environments

Introduction

Teaching mathematics to young children requires teachers to be knowledgeable in both mathematics and psychology. Equally important is a teacher's ability to develop classroom environments and events that encourage thinking and learning. Teachers need options and alternatives from which to choose. They need planning tools and resources. Four aspects of classroom teaching should be on the mind of every teacher:

1. Fitting instruction to children's learning styles
2. Evaluating, recording, and reporting children's progress
3. Arranging the children's classroom
4. Providing learning aids for children

Fitting Instruction to Children's Learning Styles

Both children and teachers have distinct styles of operation in the classroom. Each child's personal learning style is unique to that child. It determines how the child discovers relationships, learns to read, and develops the concept of number. One child may be "at home" in the classroom, comfortable with the surroundings and other children; another child may be ill-at-ease, self-conscious, and shy. As a consequence, one child may learn quickly, barely assisted by the teacher, whereas another child may learn slowly and be dependent on an adult. And when fifteen or twenty-five or thirty children are collected together for instruction, the combinations of various learning styles present a formidable challenge to a teacher. Likewise, each teacher's personal teaching style is determined by that teacher's beliefs, experiences, education, and expectations of children's behavior.

Learning and teaching must be planned. Granted, some of the best learning may be spontaneous or incidental, but for long-term, sequential learning to occur in an enriched environment, the teacher must lay the groundwork. The teacher may wish to include children in the planning, share the learning objectives with the children, and even encourage the children to lead—all are a part of the learning process. As mentioned in chapter 1, the teacher needs to be aware that many alternatives exist when developing teaching-learning strategies. Although never complete, a listing of many teaching-learning strategies follows. Each particular strategy contains these components:

1. A single major *focus*
2. A single major type of *student involvement and accompanying teacher behavior*
3. A single major type of *student grouping*

Figure 8-1 illustrates the three components of a teaching-learning strategy.

FIGURE 8-1

When designing learning environments based on sound principles of learning, the alternatives presented will suggest a variety of different and effective teaching-learning strategies. Still others will emerge as thoughtful teachers reflect on how they teach and wish to teach. Developing any particular strategy is a unique function of an individual teacher and may only be successful under conditions experienced by that individual. Although it is recognized there is no one best way to teach, teaching in only one way, that is, using one strategy continuously, is not as likely to succeed over a period of time as is using a variety of approaches. For those who have not yet taught, the following options will demonstrate the range of available teaching-learning strategies. Those who have taught will be reminded of the many possible teaching methods that exist.

Options for a Focus

The primary focus of a learning experience represents what the lesson is about. The focus is the general intent of the lesson culled from the teacher's objectives. For example, the teacher whose objective is for five-year-old children to discover different ways to sort logic blocks might have "Problem Solving" as the lesson's primary focus. Of course, if the children have never sorted materials before, the primary focus might be to "Introduce a Topic." Because there could be several simultaneous foci, it is important to determine the *primary* focus. The focus provides direction for a lesson. The following are listed alternative foci, their meanings, and classroom examples to illustrate them.

> *Focus 1:* Introduce a Topic
>
> Meaning: Introducing a mathematical topic not previously studied by the student.
>
> Example: Introducing the concept of addition to seven-year-olds.

> *Focus 2:* Reintroduce a Topic
>
> Meaning: Reintroducing a mathematical topic previously studied.
>
> Example: Using a group counting activity with four-year olds.

> *Focus 3:* Improve Attitudes
>
> Meaning: Developing favorable attitudes toward and appreciation for mathematics.
>
> Example: Employing a colorful board game to reinforce grouping skills.

> *Focus 4:* Develop Computational Skill
>
> Meaning: Developing facility to perform basic computations.
>
> Example: Involving eight-year-olds in repetitive practice to help them memorize the basic addition facts.

> *Focus 5:* Problem Solving
>
> Meaning: Solving mathematical problems and puzzles.
>
> Example: Having five-year-olds complete a pattern of cutout shapes.

> *Focus 6:* Evaluate
>
> Meaning: Evaluating the work of a child.
>
> Examples: Using a diagnostic inventory with a four-year-old.

> *Focus 7:* Enrich
>
> Meaning: Extending or deepening the mathematical understanding of a youngster.
>
> Example: Showing an animated film of shapes and their relationships to a group of seven-year-olds.

Options for Student Involvement and
Accompanying Teacher Behavior

The primary type of student involvement along with the appropriate teacher behavior are closely related in teaching. How students are to perform limits how the teacher interacts with them. For example, if a teacher would like the children to participate in a discussion, the teacher must allow and encourage a discussion to take place. This is done by leading the discussion, asking questions, answering questions with questions, giving occasional opinions, listening to students, allowing students to interact with students, and mediating between students. Lecturing would be an inappropriate technique for encouraging discussion. Again, more than one type of student involvement may be desired, so it is important to identify the primary type. The following are alternative types of student involvement, accompanying teacher behavior, and classroom examples.

Student Involvement 1: Freely Exploring and Discovering

Teacher Behavior: Observing; little or no interacting with the children.

Example: A group of children is left alone to explore and familiarize themselves with the geoboard. The teacher occasionally observes their progress.

Student Involvement 2: Directed Exploring and Discovering

Teacher Behavior: Asking leading questions; giving few or no answers; answering questions with questions; interpreting directions.

Example: A class is encouraged to discover the relationship between two different sets of materials with common characteristics, such as four red triangular shapes from the logic blocks and six red cubes. The teacher asks leading questions.

Student Involvement 3: Discussing

Teacher Behavior: Leading discussion; asking questions; answering questions with questions; giving occasional opinions; listening to students; allowing students to interact with students; mediating.

Example: Children and teacher discuss a topic together or children discuss with one another a topic such as, "Why is joining a green with a yellow Cuisenaire rod the same as joining a yellow with a green Cuisenaire rod?"

Student Involvement 4: Listening and Looking

Teacher Behavior: Lecturing; explaining; describing; demonstrating; using educational media; participating with children.

Example: Children watch and listen to one of the children's mothers explain what a computer is and how it helps her at work. The teacher listens along with the children.

Student Involvement 5: Drilling

Teacher Behavior: Administering; timing; encouraging; working with individuals; coaching; observing.

Example: Children participate by playing a card game that reinforces the basic addition and subtraction facts. The teacher observes their progress by walking to where the game is being played every now and then.

Student Involvement 6: Evaluating

Teacher Behavior: Examining; supplying data; discussing; negotiating; observing; recording.

Example: The teacher asks several questions to individual children to determine the readiness of the children to expand their knowledge of the concept of number.

Student Involvement 7: Writing

Teacher Behavior: Assigning worksheets, workbooks; dictating; observing.

Example: Children are given a worksheet on which they are to match large and small objects by drawing lines from one to the other.

Student Involvement 8: Presenting

Teacher Behavior: Listening; discussing; responding.

Example: A group of three children tell a "number story" to the rest of the class, and each time they say a particular number they clap the appropriate number of times.

Student Involvement 9: Nonverbally Communicating

Teacher Behavior: Leading; participating; observing.

Example: Nobody speaks during an activity in which the teacher illustrates a set of objects on the chalkboard, and individual children go to the board and put an appropriate number of tallies or the numeral describing the number property of the set.

Options for Student Grouping

How a particular group of children is organized for learning depends on what is being taught, the nature of the youngsters, and the mood the teacher wishes to portray. The more mathematical experiences children have, the greater the diversity of their interests and talents. It is imperative for teachers to know the alternatives available for grouping children of varying backgrounds. For example, when only one set of logic blocks is available for a class of twenty-five, the primary type of student grouping would likely be "small group." Thus, two groups of four children might each use one half of the logic blocks while the other seventeen children might be broken into small groups for using worksheets or participating in a language activity. A listing of alternative types of student grouping follows. It is quite possible to have two or more types of grouping simultaneously.

> *Grouping 1:* Small Group
> Meaning: A group of from 2 to 10 students.
> Example: Four children working with the logic blocks to develop skill in using order relations.

> *Grouping 2:* Large Group
> Meaning: A group of from 11 to 20 students.
> Example: A group of seventeen children playing the counting game Buzz.

> *Grouping 3:* Individual
> Meaning: A single child.
> Example: The teacher administers a diagnostic inventory in mathematics to one child.

> *Grouping 4:* Whole Class
> Meaning: An entire classroom of children.
> Example: The entire class views a film depicting various geometric shapes dancing about.

> *Grouping 5:* Extra Class
> Meaning: A group consisting of more than one classroom of children.
> Example: Two kindergarten classes meet together to hear a speaker discuss an upcoming field trip to the zoo.

As the teacher ponders the developing of a teaching-learning strategy, the teacher may identify one alternative from each of the three components discussed above: Focus, Student Involvement and Accompanying Teacher Behavior, and Student Grouping. Variety in teaching approaches arises as the teacher selects from the numerous combinations that exist under each component and varies the choice each time a strategy is developed. Still further possibilities will come to light during a later discussion of various arrangements of the physical classroom environment. One choice may affect another. For example, what is chosen as a focus may affect the type of student involvement and grouping.

Planning for Teaching

Excellent teaching occurs, in part, because it is well planned. Teachers are seldom able to spontaneously lead children day after day without pondering, reflecting, deciding, anticipating, and researching. Teachers must be every bit the learners they expect their children to be. Planning for teaching takes numerous forms. For the uninitiated preservice teacher, planning must be carefully prepared, often written in a detailed lesson plan, so that the teaching situation can be controlled and later analyzed by both the teacher and the cooperating teacher or supervisor. The preservice teacher gains skill and confidence in the ability to think through the teaching act when it is in a written plan. Such a plan prepares teachers in a thinking process that eventually frees the teacher from having to write out detailed lesson plans. An outline for designing a lesson plan is presented below.

Designing A Lesson Plan

I. Select the topic (concept or skill) to be taught.

II. Research the topic as follows:
   A. Discover various introductory techniques appropriate to the topic.
   B. Select teaching aids and materials which will best illustrate the topic.
   C. Identify various reintroductory methods to aid in mastery of the topic.

III. Consider how you will make this topic (concept or skill) *interesting* and *worthwhile* for both you and the children.

IV. The following may help you plan your lessons:
   A. State the overall focus for the set of lessons in terms of what the teacher hopes to accomplish.
   B. Each individual lesson will consist of the following:
      1. State a specific instructional objective for each lesson in terms of what each child should be able to do when the lesson is completed.
      2. List the mathematical terms which may need to be reviewed or learned by the child.
      3. List your learning aids and materials and explain the part they will play in the presentation of the lesson.
      4. Outline your teaching strategy (consider alternative strategies).
         a. *Introduction*—Exactly what device or technique will you use to motivate this particular lesson? How will you assure that the children will be interested?

     b. *Procedure*—How do you plan to succeed in achieving the instructional objective? Be descriptive. What will the students be doing? How will they be organized? Is this different than the way you taught them last time?

     c. *Follow-up*—Have you planned a game, problem-solving activity, braintwister, worksheet, and so forth? What is it and what is its purpose?

  5. State how you will determine if the child has achieved the behavior as stated in the instructional objective (a test is only one of many ways).

C. Evaluate your teaching performance in light of how the child responded, how you responded, and the relative success of the lesson. What might you do differently the next time you taught this lesson or group of lessons?

By the time teachers are engaged in their first years of teaching, they should be able to plan a week at a time by jotting down topics and key ideas they wish to teach. They will not need to detail every activity but may wish to write out those activities that need special planning. Much of the planning, including objectives, activities, grouping, and room arrangement can be done in the mind. Regardless, it is sound-practice to have the week's plans written out in global terms.

Experienced teachers must plan every bit as much as those less experienced. But experience has taught them what to expect, how to react, how to time a lesson, and ways to interest children. Most of the planning of experienced teachers takes place in their minds. Often, they have taught a particular topic before, maybe several times. Thus, all they will need to do is refine and prepare the teaching to fit the particular group of children with whom they are working. Experienced teachers are aware of their personal teaching styles and can adjust their styles to fit the learning styles of their children. Experienced teachers can be a bit more spontaneous and less tied to a fixed lesson plan.

Throughout the process of planning, teachers must be active learners. They must be current in what is happening in the world because the children will voice their concerns, and teachers should be ready to use children's concerns in present and future lessons. Teachers must also be current in the various subjects for which they are responsible. To be current in the teaching of mathematics, the teacher may become involved in graduate courses or in-service courses offered by colleges or school districts. They may attend local, regional, or national meetings offered by the National Council of Teachers of Mathematics (NCTM), the professional organization for those teaching mathematics at any level. Another way for teachers to keep current is to sponsor building or district mathematics fairs at which the work of children is displayed and judged.

Photo 8–1. *Arithmetic Teacher,* Oct. 1977. Reproduced by permission of the National Council of Teachers of Mathematics.

Periodicals for teachers also offer monthly ideas for teaching mathematics. Most appropriate among magazines available for those who teach mathematics to young children is the *Arithmetic Teacher,* published by the NCTM. Other magazines devoted to general classroom teaching have outstanding monthly features that present teaching ideas in mathematics. As well, the NCTM publishes yearbooks on timely mathematical topics. The 37th Yearbook, *Mathematics Learning in Early Childhood,* was devoted entirely to teaching mathematics to young children. With such a wide range of available resources, teachers of mathematics should be able to provide imaginative and up-to-date teaching experiences for their children.

Evaluating, Recording, and Reporting Children's Progress

Evaluation is a multifaceted process. It involves determining the amount and quality of children's growth, development, and achievement. It involves knowing the objectives of the mathematics program. It requires teachers to know and understand children. Evaluation also includes diagnosing, recording, and reporting children's progress. Evaluation is not an adjunct to the mathematics program. It is an integral part of daily instruction, for a teacher cannot proceed without knowing where the children are in their maturation, experience, and mathematical knowledge. This evaluation begins whenever the teacher first comes into contact with children and parents. It occurs every day throughout the school year.

Evaluating is gathering information about children using many available techniques. Recording is a way of keeping the information once gathered. Reporting is a way of sharing the data with those most concerned with children's progress—the child and parents. Each of these three aspects of the evaluative process is discussed below.

Evaluating Children's Progress

The four primary modes of collecting data about young children's progress are (1) diagnostic interview, (2) teacher observation, (3) teacher-made tests, and (4) standardized tests. The first two are particularly appropriate for young children because interviews and observations tend to be less formal and threatening than other testing procedures. The diagnostic interview takes place when the teacher sits with individual children and asks them to perform simple tasks or to comment on tasks that are performed for them. Piaget has devised many such tasks for use with young children to determine the children's readiness to learn various mathematical topics. Topics for which tasks have been devised include the concept of number, the four basic arithmetic operations, measurement, logical thought, space, time, and fractions. The Piagetian tasks and those similar to them are carefully described in books by Copeland and the Nuffield Project. Teachers who wish to discover children's developmental levels relative to mathematics are encouraged to use one or more of these diagnostic aids.

The diagnostic interview allows the teacher to employ an individualized technique in an informal setting. During the interview, children should be relaxed and comfortable. The short attention span of young children should be taken into account. Questions may be an outgrowth of children's responses; thus, teachers are able to avoid the rigidity of more formal testing procedures.

Photo 8–2

The results of the diagnostic interview may be summarized by anecdotal remarks or on a checksheet.

Experts generally agree that teacher observation is valuable, if not the most valuable method for objectively appraising the progress of children. As in the diagnostic interview, teacher observation focuses on the individual child. Certain aspects of children's behavior are not evaluated by tests. These behaviors include attitudes and interests, creative tendencies, and children's abilities to explain their own and other children's solutions. During discussions, work periods, play, and instructional time, observing teachers are mentally collecting information about individual children, and should, as soon as possible, jot their observations in the children's mathematics folder or wherever such information is kept.

Teachers gather useful information by effectively asking questions. They may ask: "Why is that the next block in the pattern?" "Can you show us how you found out?" "How do you know?" Sometimes observations involve merely watching children solve problems, such as building with blocks or bricks, drawing pictures, and shopping at a classroom store.

Photo 8–3

The observing teacher will gain information about all aspects of children's cognitive growth—language and communication, social awareness, curiosity—not merely mathematical growth. This information will help guide the teacher in planning classroom experiences in a range of subjects. Because teachers make observations every day they are with children, teachers should accept this technique as providing the most consistent and abundant source of information about children.

Teacher-made tests can be constructed to gather data about concepts and skills that have been presented in the ongoing mathematics program. These tests have a high degree of curricular validity, and for that reason, they are advantageous. Teachers can be specific regarding concepts being tested and can test immediately preceding or following the presentation of a concept to an individual, small group, or entire class. The teacher-made test is an original creation of the teacher. Young children require informal, easily understood, illustrated exercises. Older children may be tested using more symbolic tests. An example of each type is shown in Figure 8–2.

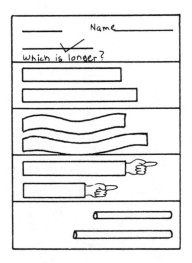
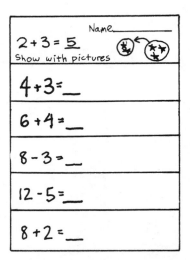

For a four- or five-year-old.   For a six- or seven-year-old.

FIGURE 8–2

As children's verbal and mathematical skills increase, teachers have a variety of testing formats from which to choose. They include simple recall and free response types of tests, the main purpose of which is to appraise vocabulary, computational skill, and understanding of number operations. There are alternate response items that test recognition of generalizations, place value, and estimation answers in mental arithmetic. There are multiple-choice items that lend themselves to evaluating vocabulary, identifying processes in problem-solving, estimating answers, checking solutions, and understanding the number system. Completion items may be used to check knowledge of number relations, understanding, and recognition of generalizations and vocabulary. There are matching items that can be used to evaluate measurement concepts and understanding of the number system. An overriding characteristic of teacher-made tests should be the practicality of the test items. That is, applications of mathematics should be stressed, not merely numerical responses.

Standardized tests are probably the most controversial of evaluative procedures. They must be used with understanding and discrimination. Standardized or norm-referenced tests are typically used for comparing the work of an individual or group with norms established for children of similar age-groups or grade levels. Sometimes standardized tests help survey children's skill and knowledge to provide bases for remedial work or grouping. Standardized tests measure speed and accuracy, some problem-solving ability, and vocabulary. All of the findings are rated and placed on a scale representing regional or national norms. Teachers should avoid relying heavily on standardized tests as a basis for evaluating children on all phases of a mathematics program. Standardized tests, when based on national norms, tend to be merely approximations of how well particular children in a particular section of the country perform. The results are always global. Such tests give little evidence of children's resourcefulness or confidence in attacking new problems. Yet, standardized tests serve a valuable function. They provide a general statement about the progress that groups of children are making. They are constantly being revised to reflect modern trends in teaching mathematics. They are prepared by experts in test construction and leaders in mathematics education. The most up-to-date tests for young children can provide information difficult to obtain by other means.

Teachers have a major obligation once data have been gathered by any of the above means. Namely, they must decide what the data mean. Teachers' interpretations of children's work are crucial in laying the groundwork for further instruction. Tallying right and wrong answers is not nearly as important as determining how the child is thinking. It has been said, "Children are not wrong; they merely respond to the stimulus according to their knowledge and development at the particular time when we are checking their progress." (Biggs and MacLean, 1969, p. 191). Teachers need to keep this in mind and to provide children the encouragement they need to maximize their potential. Young children naturally learn from failure and mistakes. All children should realize that evaluative procedures represent another natural step in the learning process.

Recording Children's Progress

During the entire evaluation process, organizing the information gathered is essential. The data must be first recorded, and then put in a place in which they are easily accessible and as easily read and interpreted. Perhaps the most useful place for storing information is a mathematics folder or combination of folder and daily workbook kept by the children. A child's mathematical folder would contain the anecdotal records and checksheets from diagnostic interviews, teachers' observations, test results from teacher-made and standardized tests, and his day-to-day workbook or worksheet samples. If each child keeps a daily workbook, his or her daily work would be kept together and occasionally reviewed by the teacher.

Throughout the year, the teacher and children can review the growth and progress the children make. When parents confer with teachers, the mathematics folder provides samples to reinforce the teacher's specific evaluative comments. Of course, there are alternatives to the mathematical folder. A notebook may be kept with pages for each individual. Other ways include using a standard gradebook or a card file system. It is very important that some systematic record keeping procedure be employed.

Reporting Children's Progress

Parents and teachers need to communicate about the children's growth and progress. The interchange between parents and teachers helps the parents determine their child's progress and helps the teacher determine the nature of

Photo 8–4

the child's home environment and the child's perception of the teacher and school. The key to two-way communication is the teacher's willingness to share whatever information and observations have been made. Teachers should welcome parents to school and feel comfortable in the children's homes. The natural interest of parents and teachers in the growth and development of children should draw both together. The teacher's assessment and parents' observations should blend together to strengthen the parent-teacher partnership.

Parents of young children are particularly interested in hearing about their children. Besides quarterly or mid-year conferences, parents may be invited to the classroom to observe the daily routine or specific activities such as a mathematics lesson. Notes may be sent to parents pointing out a significant event or accomplishment of individuals or small groups of children. A note home to a parent need not have a negative connotation. A summary of the past week's activities or of those events to come may be distributed to keep parents informed of classroom life. The help of parents may be solicited when parents are known to have particular skills or experiences that can enrich a class. Invite parents to participate.

When reporting the mathematical progress of children, written descriptions along with samples of children's work are most helpful. Sometimes a concepts-and-skills checksheet may substitute for the written description. When children have been involved in self-assessment, their views and perceptions add to the sum of information about mathematical progress. At least twice a year, and hopefully more often, teachers and parents should meet face-to-face to discuss the mathematical work and growth of children. Above all, the teacher should be open and straightforward in discussing children. The concerns of parents and teachers should be coordinated for the benefit of the children.

Children with Special Needs

In recent years, federal and state legislation has emerged that is designed to increase educational opportunity for greater numbers of children. Special programs have been mandated for children with special needs. Both slow learners and gifted youngsters will be affected as will be most classroom teachers. *Mainstreaming* will challenge the skill and sensitivity of all teachers.

Mainstreaming refers to placing exceptional children in regular classrooms for most of the school day. Thus, some children who previously spent all of their time in special education classes will now take their place in regular classrooms. By placing these children in regular classes as much as possible, the hope is that children will have a better social adjustment as well as a strong educational program. This places demands on the teacher for individualized planning. It also demands caring attitudes from both children and teachers. Teachers must help children be accepting of their new classmates.

Public Law 94–142, the Education for All Handicapped Children Act, calls for individualized diagnosis, prescription, instruction, and evaluation for all children. Teachers must plan individualized educational programs for each child. This section looks at some ways this may be done for all exceptional children—handicapped and gifted.

Children with Special Problems

Low achievers, underachievers, educationally disadvantaged, culturally deprived, emotionally disturbed, and learning disabled are all labels used to describe children having difficulties in school. Teachers are cautioned to avoid labeling children and to consider children for both their strengths and weaknesses and as full members of the human family. If a child happens to learn more slowly than others, he deserves special attention in developing early number awareness.

Slow learners may be identified by intelligence quotient, achievement, teacher observation, reading ability, Piagetian stages, and so on. They generally fall below an average in one or more of these areas. They may have many things in common, but each child is unique with his own set of strengths and weaknesses. Characteristics possessed by slow learners are listed here.

1. *Self-concept.*   Children may learn at a very young age that they are "stupid." Failure is too easily learned. Many children will not even attempt a task because they are afraid of failure. Teachers must make an effort to ensure success and to look at failure as an acceptable route towards learning. If children view themselves as worthwhile, they are more apt to approach a problem with confidence and to have a greater chance for success.

2. *Attention span.*   Slow learners often have short attention spans. This may be because problems are too difficult, too long, or uninteresting. Children will work for relatively long periods on interesting problems suited to their level. Teachers must ensure children are positively motivated toward appropriate tasks.

3. *Specific mathematics disability.*   Terms such as acalculia, number blindness, and specific minimal brain damage are used to describe children with specific problems in learning quantitative concepts. There may also be perceptual problems that affect learning spatial concepts. These include forming concepts such as position in space (near, far, up, down, left, right), distinguishing a figure from the surrounding background, and eye-hand coordination. Children are easily distracted by extraneous stimuli. Too many problems or pictures on a workbook page, as well as too many objects or people in the classroom, can be very distracting. Workbook pages and classroom environment may need to be relatively plain and simple for children with perceptual problems. Teachers should be alerted to these problems and should seek professional help when problems demand it.

4. Poor self-control. Some children may be explosive, hyperactive, or erratic. They always seem to be in motion. They rarely sit still and often wander aimlessly about the room. Some research indicates that these actions may be triggered or aggravated by diet. Much research still needs to be done on how diet affects children. Hyperactive children require a structured environment with few extraneous distractions.

5. Brain dominance. Recently, attention has focused on the role played by the left and right hemispheres of the brain. The left hemisphere often seems to affect verbal, numerical, and logical functions; the right hemisphere seems to affect visual, spatial, perceptual, intuitive, and imagination functions. Teachers often direct lessons to the left hemisphere. Children are asked to read, listen, think, and write. This may be suited to children dominated by the left hemisphere, but causes problems for the right hemisphere children who need to see, feel, manipulate, and imagine. Most research in this field recommends strengthening both hemispheres of the brain and the bonds between them. The right hemisphere of the brain seems to be neglected in Western cultures, and teachers can do much to strengthen it by creating opportunities to visualize and imagine.

6. Language problems. Children who have difficulty learning mathematics often have language difficulties. They may not understand such common mathematical vocabulary as *up, down, in, out, two, plus,* for example. They may be unable to read simple directions, equations, or mathematical symbols. They may be unable to communicate concepts they do understand. Teachers should remember to keep their conversations as simple as possible and be alert for any misunderstanding of terms. Concepts should be developed through physical manipulation and language.

7. Memory and application. Studies show that slow and retarded children are capable of learning complex motor and verbal skills. Their retention may be similar to that of younger children of the same mental age. Overlearning may be required to ensure retention. Teachers need to allow for practice, drill, and repetition but only after a concrete understanding of concepts has been developed. Transfer of learning is difficult but may be accomplished if it is incorporated into the lesson. Slow and retarded children can retain and apply skills when transfer has been practiced. Complex problem solving may be too difficult; but simple, rote, factual material such as the memorization and use of basic facts can and should be learned and applied.

8. Piagetian stages. Some children may be slow in mathematics because they are in an earlier Piagetian stage than most of their age-mates. Young children may be in the sensori-motor stage, while their classmates are in the preoperational stage. Second or third graders may be in the preoperational stage, while others are entering the concrete operational stage. Topics must be presented differently to these children. They may not be ready for some concepts.

For example, a preoperational child who does not conserve number is not ready to learn the numbers from one to ten. This child needs additional pre-number work.

General principles of good teaching have been mentioned throughout the book and pertain to all children. They are reiterated here because of their significance for children with special problems.

1.   All children are ready to learn something. The teacher must determine the exact level of that readiness.

2.   Success is important. Learning must be carefully sequenced to ensure several levels of success. Immediate positive feedback is helpful.

3.   Self-concept affects success and vice versa. Children must be worthwhile in their own eyes and in the eyes of their peers.

4.   Practice is important and should follow the concrete development of concepts. It should be applied in practical situations and should contain provisions for transfer.

5.   The teacher needs to be prepared with a variety of teaching strategies. These should not all be presented at once; but if one method fails, another must be tried. The child might be capable of mastering a concept, but not in the context in which it was first presented. Several methods of presentation that involve the senses may be needed to meet each child's individual learning style.

6.   Children with problems in mathematics may work slowly and at a concrete level. They may be incapable of abstract work. More time may be needed on concrete levels of development.

7.   Children's mistakes should be analyzed carefully. Children periodically make careless mistakes, but there is often a reason for the problem. Teachers need to look for patterns in children's errors, to discuss their thinking processes, and to correct the mistaken concepts.

8.   Learning is different for each child. This chapter mentions several ways to diagnose and evaluate individual children. Lessons should be planned according to this diagnosis.

Children with Special Mathematical
Talents

Just as children having problems in mathematics need special consideration, so do children with special talents in mathematics. The regular mathematics curriculum may be as unsuited for gifted children as for slow learners. The right of children to an education suited to their individual needs has prompted some states to legislate special programs for gifted children. Whether gifted children are in a regular classroom or a special program, there are several things to consider.

Gifted children, of course, are not all alike. Although they are commonly identified by high intelligence quotients, such as a score of 130 or better, there are several other characteristics to consider. Some of these are listed here.

1. *Creativity.* Many children, from preschool on, will have special creative, mathematical talents. Children should be encouraged to explore, manipulate, and suggest a variety of solutions to a problem. Rote memory should be minimized. Flexible thinking should be stressed.

2. *Awareness.* Gifted children are often sensitive to and aware of their surroundings. They perceive problems readily and can see patterns and relationships easily. They do not need to have every step of a problem spelled out for them.

3. *Nonmathematical abilities.* Although some children will have special talents only in mathematics, many young children who are mathematically talented are also physically stronger and more attractive, mentally and emotionally mature, and highly verbal. They can express their thought processes.

4. *Abstract reasoning.* Talented children are generally able to reason at a higher level of abstraction than their age-mates. They can work symbolically with quantitative ideas.

5. *Transfer.* Skills learned in one situation can be transferred to a novel, untaught situation by a gifted child. Learnings may be applied in social situations, other subject areas, home problems, and so on. Generalization of rules and principles should be encouraged and tested in new settings.

6. *Memory.* Talented children have the ability to remember and retain what has been learned. They do not need as much drill as other children.

7. *Curiosity.* Mathematically gifted children often display intellectual curiosity. They are reluctant to believe something just because the teacher says it is true. They want to know why it works. They are interested in a wide range of ideas and will often explore topics through independent reading. They ask many questions, enjoy solving puzzles, and delight in discovering winning strategies in games. Challenging articles, books, puzzles, and games should be available for their use.

Teaching the talented and creative child is challenging and even threatening to some teachers. They are sometimes concerned that they cannot answer all the childrens' questions or that children may be smarter than they are. This may be true but should not necessarily be a problem. Some suggested techniques that are especially appropriate for gifted children follow.

1. Gifted children need challenging problems. Several of the activities suggested in chapter 7 are especially appropriate for gifted children. They enjoy strategy games and complex problems.

2. Gifted children do not need *busy work*. Because they often finish assignments early, they may be asked to do additional problems of the same type. If they can do the initial problems, it is likely they do not need more practice. Let them move on to more challenging tasks.

3. Encourage independent research. Gifted children are often very capable of independent study and research at a young age. They may even set their own goals and evaluate their progress. This, of course, does not mean they should always be left alone.

4. Set up a *buddy system*. Let children work with peers of similar ability, or let them help others who may be having problems. Children can often communicate with and learn from peers better than from adults.

5. Encourage creative and critical thinking. Avoid forcing children to memorize one method to the exclusion of others. Accept any correct method, and lead children to discover why some methods will not always work. Children should evaluate solutions for appropriateness, ease, originality, and so on.

Educators often debate whether programs for gifted children should be enriched or accelerated. Some states have mandated one type of program to follow. In early childhood education, the trend seems to be toward enrichment or a combination of enrichment and acceleration. Enrichment includes introducing topics not normally considered as part of the curriculum for young children and exploring the more traditional topics in more depth. This requires the availability of materials other than textbook and workbooks. The teacher may construct games, task cards, and other materials.

A wide range of source books should be available either in the classroom or the library. For the teacher, a professional library, the National Council of Teachers of Mathematics, journals, and activity books are rich sources of good ideas.

Arranging the Children's Classroom

Classrooms are for children. Classrooms should provide the most expedient physical environment possible for learning. It is possible to have a classroom too stimulating, with so many bright pictures and objects that children have difficulty finding a single object or area in which to play that is not distracting. It is possible to have a classroom so barren that children have little or nothing in which to get interested. It is possible to have a classroom so informal that children do not know what to do or where to go. It is possible to have a classroom so formal that children become regimented and repressed. It is a challenge to teachers to provide a balanced, comfortable, flexible classroom, so that no matter what activity has been planned, there is space available that lends itself to achieving the focus of the activity. Also to be considered are the learning styles of children and teaching styles of teachers.

Several basic tenets should be kept in mind in providing the physical learning environment.

1. The physical learning environment provides a support system for the educative process. The stimulation for learning emerges as much from the physical environment as from the textbooks, teachers, games, and media hardware.

2. As learning environments become better, so also the learning becomes better. That is, there will be improvement in mathematical concept and skill development as well as better communication.

3. Full use of the learning environment will assist the teacher in teaching every child more fully. Children will receive help from the teacher and the environment.

4. Classrooms for children should mirror the decisions and interests of the children. Children should have the opportunity to help design parts of their own learning environments. Teachers should also help design the learning space.

5. Children's behavior is affected by their learning environments. Anything that affects behavior also affects learning. The quality of the physical environment must be maintained.

6. Teachers need to be aware of the physical environment as part of the learning process. Alternative ways to use space is fundamental for this awareness.

7. The transformation of a classroom relies on the awareness of those who spend time in that classroom. Children and teachers do not need a new building or a new classroom to have a rich learning environment.

8. Use of the natural environment in design has served the architectural community well. Analogous to using the natural environment is using the human and physical needs of children in designing learning environments. Taylor and Vlastos underscore this belief: "If the child's experiences are a starting point for education, then, by definition, his own language and the culture of his home, neighborhood, and community should be utilized in the educational environment.

The classroom and outdoor area should play a vital role in reflecting a child's cultural background and his interests." (Taylor and Vlastos, 1975, p. 24).

From these tenets comes the key to developing more effective learning environments: that of *adapting* existing classroom space. Aspects of the children's world and their learning of mathematics as described in chapter 1 are that children: (1) want number experiences, (2) are active, (3) constantly observe relationships, (4) live in a multifaceted world, (5) rely on teachers to help develop learning environments, (6) are expected to learn mathematics and (7) have feelings that need consideration. These characteristics of young children provide the starting point in determining how a specific classroom should be structured.

Children constantly use their senses. Thus, the visual stimulation must be active and must help develop relationships. Paint and graphics on the walls draw attention to the cultural backgrounds of the children. Children as artists and planners can provide a pleasing, human, child-oriented environment. A "wall of numerals" (as in Figure 8–3) can replace the unstimulating numerals over the chalkboard.

21	22			▮	23		▮	24	25	
▮		▮	16	17		18	▮		19	20
	12	13	▮		▮		14	15	▮	
6	▮	7	8	▮	9	10		▮		11
	▮	1	2	▮		3	4	5		

FIGURE 8–3

A corner of color featuring a different color each month with objects of that color, variations of the color, children's experimentation with the color, and so forth interests students. The final result should be displayed within the children's scale of vision, not in an adult's scale of vision, that may be visually discarded by children.

Auditory stimulation may be provided by record or tape players or by the school intercommunication system. Often, the only sound reaching the children's ears is that of the teacher's voice. Communication skills may be more appropriately ordered by periods of active listening and speaking by the children. Listening to music, stories, poetry, and sounds as well as recording their own voices while sharing their writing is a stimulating exercise for children. Mathematical awareness can be enhanced by listening to sound patterns, number stories, and descriptions of relationships. On the other hand, children enjoy recording their own "Story of Three" or a clapping pattern to be completed by other children.

Likewise, classification by tasting, smell, and touching may excite the senses to gather information from the environment that makes mathematics more meaningful. Sand and water play simultaneously builds the foundations for measurement and provides rich experiences in touching and comparing. The classroom that foregoes children's sensory learning potential is providing a sterile learning environment.

When a teacher considers personal needs, those of the children, and those of the curriculum while understanding the need for a stimulating learning environment, an adaptable experience-centered classroom will likely emerge. Teaching will change. The process of learning will change.

Whatever classroom is inherited by the teacher, much can be done to provide a rich learning environment. A complete transformation is not expected immediately. There are limits imposed by the classroom structure, time, money, and other school personnel. Awareness of the existing potential of each classroom and sensitivity to the children who will spend so many hours in the classroom should serve the teacher in developing, maintaining, and changing the learning environment.

Providing Learning Aids for Children

Effective teaching is difficult, partly because it relies on a teacher's being more than a pusher of papers—from workbook pages to dittoed worksheets to the textbook. When concepts are to be learned, manipulative materials are needed. When skills are to be sharpened or facts are to be memorized, repetitive games or activities are needed. When independent work is prescribed, activity cards or learning centers are needed. Providing quality learning resources to support a sound program of premathematics and mathematics instruction is basic to effective teaching.

Acquiring learning aids depends on three factors: (1) the financial resources of a school or district; (2) the energy and creativity of teachers; and (3) the priorities established by a school or district and its intent in providing a strong mathematics program. The latter factor is influenced by the other two. Knowledgeable teachers and curriculum specialists should serve on advisory committees for developing school- or district-wide guidelines for mathematical instruction. They should recommend the wise expenditure of financial resources to provide the school and classrooms with dynamic learning aids.

Commercial Materials

Excellent commercial materials are available for use in teaching mathematics in the preschool and early years of school. They include textbooks and workbooks, kits, games, structured material, and activity cards. Each will be briefly discussed.

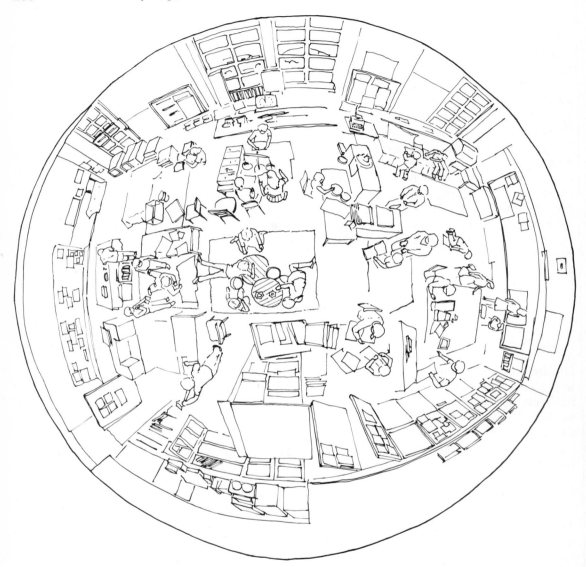

FIGURE 8–4.   A bird's-eye view of an experience-centered classroom in a traditional setting vividly illustrates the many activities that can take place simultaneously. (Taylor and Vlastos, 1975, p. 43)

Textbooks and Workbooks.   In most mathematics programs from kindergarten through high school, textbooks or workbooks provide the foundation on which the program rests. These textual materials are carefully prepared by recognized authorities in mathematics education. The materials present a uni-

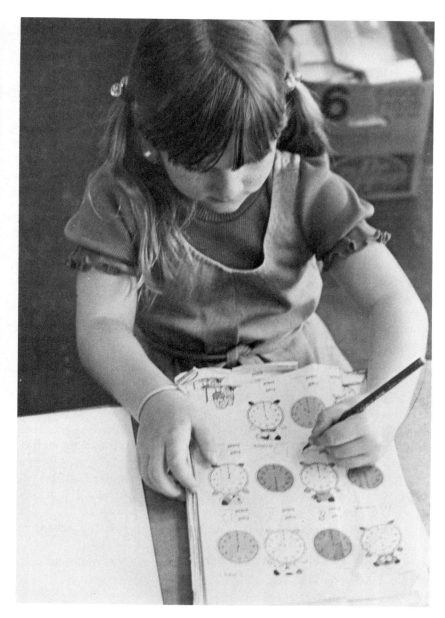

Photo 8–5

fied sequence of concepts and topics, which are reintroduced at appropriate intervals throughout the books. Textbooks and workbooks tend to be attractive,
colorful, and appealing.

Selecting the mathematics textbook or workbook series for young children is an important task, usually performed by a school district or building textbook committee. Most often, teachers have a choice of series from which to choose. Several general criteria for selecting textual materials should be considered. One such listing of criteria is presented below. A final list of criteria is the responsibility of those who actually choose the mathematics series for a district or school. It is these individuals who have considered the goals of the early mathematics program, local priorities, budgetary limitations, and teacher resources.

Selected General Criteria
for Choosing Mathematics Textbooks or Workbooks

Does the textbook or workbook:

1. Encourage active student involvement and investigation and discovery of mathematical ideas?
2. Present task-oriented problems at the student's level of understanding?
3. Suggest the use of manipulative materials?
4. Use correct vocabulary, yet avoid wordiness and undue stress that might interfere with the student's learning?
5. Provide adequate exercises to assist in introducing, reinforcing, diagnosing, and reintroducing mathematical concepts and skills?
6. Spiral the mathematical ideas so students confront an idea several times in the early years, each time at a slightly more advanced level?
7. Have an accompanying teachers' edition with valuable suggestions for introducing, reinforcing, diagnosing, and reintroducing mathematical concepts and skills?
8. Relate mathematical concepts that have common parts; for example, ordering relations with objects ("is taller than"), numbers ("is greater than"), and measurement ("is longer than")?
9. Build mathematical concepts and skills on previously understood concepts and skills?
10. Support the learning of basic addition, subtraction, multiplication, and division facts?
11. Allow for students to progress at different rates reflecting individual differences among children?
12. Interest children by being attractive, colorful, and motivating, page after page?

Kits. Commercial kits for use in mathematics learning are typically of three different types. Kits of diagnostic and learning materials generally contain manipulative materials used to test youngsters relative to their development of prenumber skills; for example, classifying, ordering, corresponding, conservation, and counting. As well, these kits may provide materials needed to develop initial understanding of classifying, relationships, number, and measurement.

Kits designed to accompany textbook or workbook series provide manipulative materials illustrated or suggested by the teachers' edition of the series. These kits may contain attribute materials, colored rods, counters, or measuring apparatus. This type of kit is a valuable resource, since it allows children to work with concrete materials as they learn mathematical ideas.

The third type of kit is a drill kit. The drill kit provides audio cassettes or cards containing basic facts to reinforce children's computational skills. The advantage to such materials is that they allow children to work independently or in small groups at the children's own level, freeing the teacher to work with other children.

When choosing a kit for classroom use, teachers should carefully review the goals of the mathematics program. The amount of time devoted to kit materials is determined by the nature of the mathematics program and the particular kit being considered. The convenience of a kit must be weighed against the amount of interest the kit will generate and maintain. If the kit can provide a function unavailable by other components of a mathematics program or by classroom materials, it should be considered for possible purchase.

Games. There is a wide selection of games available that reinforce mathematical skills. Card games, race board games, tile games, target games, dice games, table games, word and picture games, and games of probability—all may help develop skills in recognition of characteristics, counting, recognition of numerals and number patterns, probability, matching, developing strategies, and problem solving. Children are usually motivated by commercial games. Often, the children are unaware of the mathematical value of games. At times, it may be appropriate for the teacher to explain the relationship of a game to a particular skill the children may be learning in a nongame context. For example, the popular card game *Old Maid* helps develop recognition and matching. Once children have played this game, the teacher may help develop the transfer of matching playing cards to matching attribute blocks, Parquetry blocks, or tessellation patterns. The teacher should be careful, however, not to destroy the fun and motivation of the game for the sake of the mathematics.

Games for classrooms are available from department and toy stores. Teachers should carefully select games that fit clearly in the context of the total mathematics program. Sometimes the expense of games may prohibit their purchase. In such cases, teachers are encouraged to construct their own games using materials available in the school. A discussion of constructing homemade games is presented shortly.

Structural Materials. Structural learning aids are usually designed to help teach particular mathematical concepts and clearly illustrate the concept for which they were developed. Examples of structural materials are Cuisenaire rods, Multibase Arithmetic Blocks, and Logic Blocks. They, in turn, help develop number concepts and properties, place value, sorting, and logic. A more complete description of specific structural learning aids appears in the Appendix.

Although sets of structural materials tend to be expensive, they are often one of the most valuable purchases teachers or schools will make to support a sound mathematics program. Teachers are encouraged to shop around for structural materials and only purchase materials when the advantages, disadvantages, and uses of the materials are known. It is sometimes possible for a sales representative or educational consultant to offer workshops in using certain materials. To take advantage of the full value of materials, such workshops are necessary.

Activity Cards.   Activity cards are designed to lead children through developmental sequences that provide guided discovery. Activity cards have the advantage of encouraging independent work by children, thus freeing the teacher to work with other children. There are cards for use with attribute blocks, geoboards, Pattern Blocks, Geoblocks, problem solving, measuring, color cubes, and other manipulative materials. The key in considering activity cards for young children is the clarity of the instructions. Pictures and diagrams often replace words to direct children to specific activities. A set of activity cards that children are unable to understand represent a waste of valuable funds.

When considering purchasing activity cards, it is helpful to consider certain general characteristics that the cards might possess. Characteristics of good activity cards are listed below.

1. The questions or activities on an activity card should tend to be open-ended; that is, should encourage children to provide any number of responses.
2. The objectives of an activity card should be clear to the child and teacher.
3. The wording or directions should be concise and presented at the children's independent reading level.
4. An activity card should provide some way for children to record their answers or responses.
5. Activity cards should allow for a higher level thinking process than just memory.
6. Activity cards should be attractive.
7. The cards should make use of the environment—the classroom or the outside.
8. The cards should encourage active exploration and manipulation of materials (Barson, 1978, p. 53).

Teacher-Made Learning Aids

Teachers may often have more time than money or may need to tailor instruction to the needs of children. In such cases, teachers may wish to construct their own learning aids. All that is required is an interest in such a task and the willingness to devote the time and energy to do a quality job. Effective teachers are renowned for these characteristics. What follows are some examples of games or activities intended to assist children in practicing and remembering basic facts in addition, subtraction, multiplication, and division. The significance of the games described rests in the adaptability of a single game idea to many useful games that use the same strategy.

Adapting Rules or Action of an Existing Game. Most games teachers invent are adaptations of existing games they have seen or played. The simplest way to "invent" a game is to modify the rules or action of one already known. For example, the addition game Top It (p. 304), which is an adaptation of a popular children's card game called War, may be considered merely an addition game. A little reflection reveals much greater potential for practice activities.

1. If addition is to continue being stressed, the cards could be rewritten with series of three or perhaps four addends. Thus, cards might appear in the format in Figure 8–5:

2 + 5 + 3		4 + 5 + 1		5 + 1 + 6		3 + 3 + 2

FIGURE 8–5

2. As children become ready for introductory multiplication work, the addition cards might have three, four, or five addends with the same value (see Figure 8–6).

FIGURE 8–6

3. The original instructions do not mention subtraction. As addition and subtraction may be learned in concert, subtraction Top It is appropriate (see Figure 8–7):

FIGURE 8–7

Top It
(for two players)

1. Begin play with the stack of cards face down.
2. First player turns over a card and places it face up in front of him or her, giving the sum. The second player does the same.
3. The one who has the highest sum claims both cards. If the sums are the same, cards remain in front of the players and each draws another card. The player with the highest sum takes all the cards.
4. The winner is the one who has the most cards at the end of the game.

Easy Cards

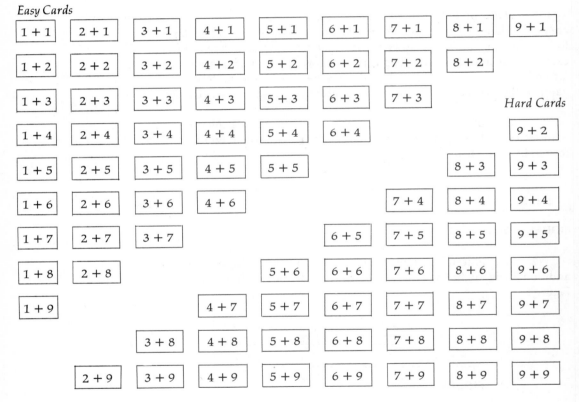

Hard Cards

FIGURE 8–8

4. Perhaps cards with mixed addition and subtraction could be used as practice for both operations. The cards might look like those in Figure 8–9.

| $4+3-2$ | $9+8-4$ | $3+3-1$ | $7+3-5$ |

FIGURE 8–9

5.   Multiplication Top It is the next variation, assisting the child memorizing the basic multiplication facts (see Figure 8–10).

FIGURE 8–10

6.   For each of the basic multiplication facts there is an associated basic division fact. The cards for this adaptation would appear similar to those in Figure 8–11.

FIGURE 8–11

7.   If there is a need for younger children to reinforce the *more than* and *less than* relationships, Top It can be constructed to assist (see Figure 8–12).

FIGURE 8–12

8.   Another way to vary the domino pattern above is to use numerals as in Figure 8–13.

FIGURE 8–13

9.   Shapes may be used on the card with the rule that a big shape "takes" a smaller shape (see Figure 8–14).

FIGURE 8–14

The essential action has been retained in each variation in Figure 8–14; that is, two cards are drawn, and the larger number (or figure) takes the smaller. In the case of a tie, another card is picked by both players.

The second way to modify a game is to change the rules. One general example will suffice to illustrate changing the rules. Each game discussed above could be constructed so the numerals or figures are one of two colors, blue or orange. The rules could be altered, so that when both players draw the same color numeral or figure the larger takes the smaller, but when both players draw different colors the smaller takes the larger. Thus, with one rule change, the games assume a different character, though maintaining the goal of providing practice for basic mathematical skills.

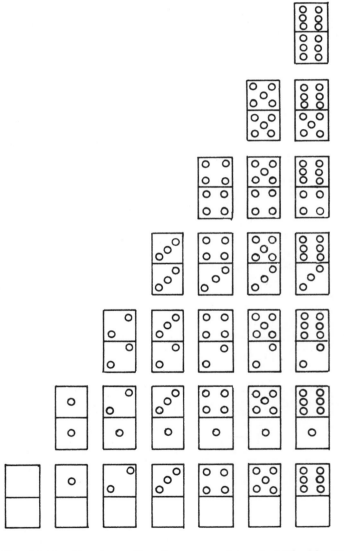

FIGURE 8–15. Making Dominoes. Here is one complete set of "double six" dominoes. It contains 28 dominoes.

The teacher is cautioned to take care in preventing boredom by spacing the use of variations of the same game and by using only those variations appropriate for the needed practice. The multitude of examples listed show how games are invented by modifying the action or rules of existing games.

Adapting Rules and Action. There are times when simultaneously adapting the rules and action of a game produces an activity quite unique from the original source of the idea. For example, consider three activities involving the creative use of dominoes. First, some description of the dominoes and their construction is appropriate. Although it is quite all right to use commercially prepared dominoes for these activities, children are particularly attracted by the larger sets that are easily constructed from stiff art board.

Domino Activity 1: Sum

Object: To score as many points as possible in a single round by summing the number of dots on dominoes.

Materials: Dominoes, paper, pencil.

Players: 2 to 5.

Play:

1. Spread the dominoes face down.
2. Each player picks one domino, turns it over, and sums the number of dots. The player with the highest number plays first.
3. Each player then chooses any 5 of the unexposed dominoes, but does not look at them.
4. The first player turns over each of his dominoes one at a time. As each domino is turned, the player sums the number of dots, announces the total, and records that amount. The first player continues to turn his dominoes, announce the sums, and record each amount.
5. When the first player has finished with all 5 dominoes, he determines the score for that round by adding the five numbers he has recorded.
6. The next player continues in the same manner, hoping to find a larger total.
7. The player with the highest score wins that particular game. Play continues as before with the opportunity for different children to win the game. Sample play for one player in one round is shown in Figure 8–16, p. 308.

Dominoes Drawn	*Domino Score*	*Announced Value*
	3 + 4	7
	6 + 0	6
	1 + 3	4
	5 + 2	7
	1 + 1	2

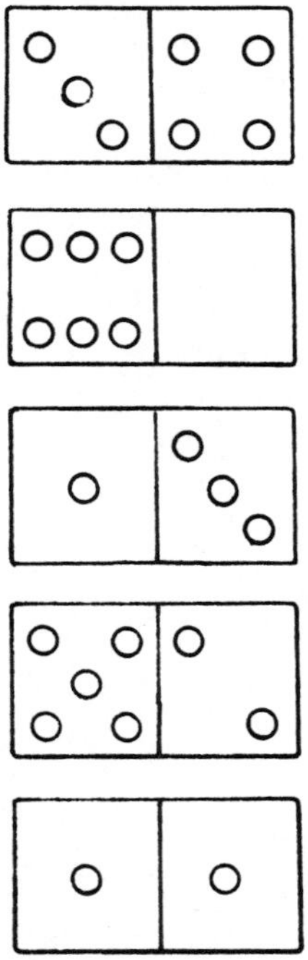

Total for round: 26

FIGURE 8–16

Domino Activity 2: Going Down

Object: To get as close as possible to zero without going below it.

Materials: Dominoes, paper, pencil.

Players: 2 to 5.

Play:

1.   Spread the dominoes face down.

2.   Each player picks one domino, turns it over, and finds the difference between the number of dots on each half. The player with the highest difference plays first (see Figure 8–17). Player 2 would play first.

3.   Each player then chooses any 5 of the unexposed dominoes, but does not look at them. The player must use exactly 4 of the 5 dominoes during the round.

4.   At the start of play, each player has 10 points.

5.   The first player turns over each of his dominoes one at a time. As each domino is turned, the player finds the difference between the numbers on each half. The player then subtracts that amount from the 10 points he started with and records his score. Because he may only use 4 of 5 dominoes, the player must decide which one to dis-

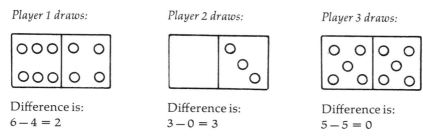

Player 1 draws: *Player 2 draws:*   *Player 3 draws:*

Difference is:
6 − 4 = 2

Difference is:
3 − 0 = 3

Difference is:
5 − 5 = 0

FIGURE 8–17

card. This decision must be made when the difference of each domino is determined. Once a domino is used, it may not later be discarded. Likewise, once a domino has been discarded all subsequent dominoes must be played.

6. The first player continues until all five dominoes have been used, or until his score goes below zero, in which case he loses the round. Play continues until all players have played.

7. The player closest to zero without going below it wins the round. Sample play for one player in one round is shown in Figure 8–18.

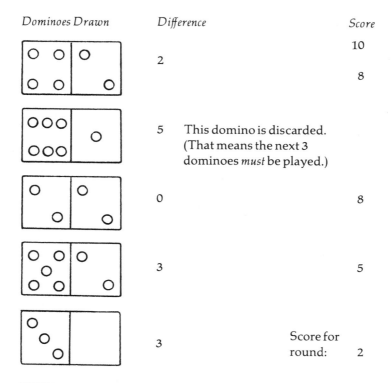

Dominoes Drawn	Difference		Score
			10
	2		
			8
	5	This domino is discarded. (That means the next 3 dominoes *must* be played.)	
	0		8
	3		5
	3		Score for round: 2

FIGURE 8–18

Domino Activity 3: Match

Object: To collect as many dominoes as possible.
Materials: Dominoes.
Players: 2.
Play:

1. Shuffle the dominoes and stack them face down in the center of the playing area.
2. Each player picks one domino, turns it over, and sums the number of dots. The player with the highest number plays first.
3. The first player declares "match" and picks a domino from the stack and turns it face up. The second player then picks a domino from the stack and turns it face up.
4. If there is a match, the player declaring "match" gets to pick up both dominoes to keep. If no match occurs, the second player picks up both dominoes. A match occurs when one set of dots (either end of the domino) of one of the dominoes matches one set of dots (either end) of the other domino, *or* the sum of dots on one domino matches the sum of dots from the other domino.
5. Play continues with the second player declaring "match." Both players then turn over dominoes to see who will keep the dominoes.
6. Play continues, with players alternately declaring "match," until all the dominoes have been taken.
7. The player with the greatest number of dominoes wins the game.

Photo 8–6

Photo 8–7

Constructing Learning Aids.

Once a sound, usable learning aid has been decided upon, construction may begin. Teachers willing to spend the time and effort to produce a long-lasting, durable aid will be rewarded by time saved repairing or remaking the activity. Game boards and cards should be attractive and colorful. Young children enjoy bright, cheery materials with which to play. These can be made with brightly colored posterboard or colorful lettering and drawings on white posterboard. Sometimes, pictures cut from magazines, ready-made stickers, or children's drawings add an extra touch of color that attracts children.

Materials should be durable. Most teacher-made materials may be protected by covering the board or cards with either plastic laminate or clear contact paper. Both protective materials are readily available at art, hardware, or variety stores. Once covered, front and back, a learning aid will last for months or even years.

The quality of an aid is improved if care is taken when lettering or attaching pictures to the material. Letters and numerals should be easily read. They may be affixed by hand or stenciled. Gummed letters or numerals may also be used. Children find it easier to participate in a game when the words, letters, or numerals are clearly visible and attractive.

Some activities need instructions. The instructions need to be clear and concise. An activity with poorly explained rules is of little value. Learning aids for young children must sometimes be explained by the teacher. The instructions the teacher reads should be carefully worded. Pictures or drawings are often useful in explaining the action of a game.

Besides motivating children, learning aids should help reinforce or teach a mathematical concept or prenumber concept. There should be no question in the teacher's mind about what concept or skill is being presented by a learning aid. As well, the teacher should see the potential in nearly every aid for its adaptation to another learning aid to teach mathematics, communication skills, social studies, science, and so on.

For example, memory (concentration) is one of the most versatile types of games. Sixteen to forty-eight cards may be constructed to reinforce shapes, number patterns, numeral-number recognition, addition, subtraction, multiplication, division, words with beginning sounds, homonyms, names and faces, animals and habitats, and so on. For young children, fewer cards (8 to 16) are used and the cards are laid out into two rows. The simplicity of memory and its easy adaptability makes memory one of the most dynamic types of learning aids.

Audio-Visual Materials

The potential of audio-visual equipment and materials should not be overlooked. There are rich resources of 16 mm films, 35 mm slides, filmstrips, cartridge films, audio and video tapes, and cartridges. These commercially available materials are easily accessible and may be found in audio-visual catalogs produced by school districts or manufacturers. Young children may engage in valuable learning experiences by developing their own learning aids. They are able to construct slides to illustrate various simple mathematical ideas. For example, sequences showing numerals and groups of objects may be developed. Slide mounts and clear acetate are readily available from photography shops. Children may develop audio tapes to describe a mathematics activity in which they have engaged. The children may tell a number story that they make up that describes "Where Two Came From" or "How Seven Got Its Name."

Equipment with which to project slides and films, along with tape recorders and listening stations, are usually available in today's schools. Also available are overhead projectors, television sets, and radios. Such a wealth for providing vicarious learning experiences should be built into teaching plans.

Other audio-visual materials include charts, models, and pictures. Their role is clear. They support the daily program of mathematics instruction. Because most audio-visual materials are one step removed from manipulation of concrete materials, children should be prepared to use such materials. A film may effectively be used to reinforce an idea previously learned concretely, to motivate further work with concrete material, to enrich the children's backgrounds, or to provide for enjoyment of mathematical ideas.

Extending Yourself

1. Select one of the mathematical topics discussed in an earlier chapter. Design a lesson plan using the outline presented in this chapter to present the mathematical topic to a small group of six- and seven-year-olds.

2. Obtain and read through a copy of the *Arithmetic Teacher* magazine. Note how some articles present activities for various levels of youngsters, whereas other articles help teachers prepare themselves for teaching mathematics.

3. Make a sketch of an empty classroom regardless of its shape. Within the confines of this classroom, design a learning environment in which children would enjoy learning and you would feel comfortable teaching.

4. Borrow a mathematics textbook or workbook for kindergarten or the primary grades. Use the Selected General Criteria for Choosing Mathematics Textbooks or Workbooks presented in this chapter to determine how you would evaluate your textbook or workbook.

5. Construct one of the games suggested for developing computational skills. Try it out with some children. Note the thinking processes of the children, while they play the game.

6. It has been said that the techniques used in teaching children with learning difficulties are the same as those used with all children in a sound program of mathematics education. Defend or criticize this statement.

7. What educational needs do slow learners and gifted students have in common? What needs are unique for slow learners? What needs are unique for gifted students. How would you organize to teach all students?

8. Consult the *Arithmetic Teacher*; 27th NCTM Yearbook, *Enrichment Mathematics for the Grades*; the 35th NCTM Yearbook, *The Slow Learner in Mathematics*, and other professional journals for articles on slow learners and gifted students. Discuss your findings with others who have read the same or similar articles.

9. Make a collection of materials that could be used with slow learners. Collect materials that would stimulate gifted children, and try them out.

Bibliography

Barson, Alan. "Task Cards," *Arithmetic Teacher*, Vol. 26, No. 2 (October, 1978), p. 53–54.

Bell, Frederick H. *Teaching and Learning Mathematics*. Dubuque, Iowa: William C. Brown Co., 1978.

Biggs, Edith E., and MacLean, James R. *Freedom to Learn*. Reading, Massachusetts: Addison-Wesley (Canada), Ltd., 1969.

Copeland, Richard W. *Diagnostic and Learning Activities in Mathematics for Children*. New York: Macmillan Publishing Co., Inc., 1974.

Design Group. *The Way to Play*. New York: Puddington Press, Ltd., 1975.

Grossnickle, Foster E., Brueckner, Leo J., and Reckzeh, John. *Discovering Meaning in Elementary School Mathematics*. New York: Holt, Rinehart and Winston, Inc., 1968.

Kohl, Herbert R. *Math, Writing and Games*. New York: The New York Book Review, 1974.

Moomaw, Vera, et al. *Expanded Mathematics Grades 4–5–6*. Eugene, Oregon: School District Number 4, Instruction Department, 1967.

National Council of Teachers of Mathematics. *Enrichment Mathematics for the Grades*. Twenty-seventh Yearbook. Washington, D. C.: National Council of Teachers of Mathematics, 1963.

──────────. *The Slow Learner in Mathematics*, Thirty-fifth Yearbook. Washington, D. C.: National Council of Teachers of Mathematics, 1972.

Nuffield Mathematics Project. *Checking Up I*. New York: John Wiley and Sons, Inc., 1970.

──────────. *Checking Up II*. New York: John Wiley and Sons, Inc., 1972.

Payne, Joseph N., ed. *Mathematics Learning in Early Childhood*, Thirty-seventh Yearbook. Reston, Virginia: National Council of Teachers of Mathematics, 1975.

Peterson, Daniel. *Functional Mathematics for the Mentally Retarded*. Columbus, Ohio: Charles E. Merrill Publishing Co., 1973.

Schminke, C. W., Maertens, Norbert, and Arnold, William. *Teaching the Child Mathematics*. New York: Holt, Rinehart and Winston, 1978.

Silberman, Charles E. *The Open Classroom Reader*. New York: Random House, 1973.

Spitzer, Herbert. *The Teaching of Arithmetic*. Boston: Houghton-Miflin Co., 1961.

Taylor, Ann P., and Vlastos, George. *School Zone: Learning Environments for Children*. New York: Van Nostrand Reinhold Co., 1975.

Thorpe, Cleata B. *Teaching Elementary Arithmetic*. New York: Harper Brothers, 1962.

A P P E N D I X

Instructional Aids and Materials

Psychologists and educators agree that young children respond best when they are taught with concrete examples of the concepts to be learned. Since mathematics is, by its very nature, abstract, it is necessary that educators "unabstract" the mathematical concepts for the young child.

Numerous instructional aids and devices have been constructed to represent various mathematical concepts. Some may be easily obtained from publishing companies and manufacturers who specialize in materials for children. Others are easily home-made.

Whatever its source, no single device thoroughly embodies a complete mathematical abstraction. In other words, if a teacher wishes to communicate to a child the idea of place value in numeration, then the child should see the multibase blocks, the chip trading activities, the Cuisenaire rods, the abacus, and also do some grouping activities with bunches of tongue depressors. The larger the number of concrete embodiments that a child sees, the more complete will become his perception of the mathematical abstraction being taught. As the child's knowledge of the concept increases, higher levels of abstraction may be used to communicate a more complete picture of the idea.

This appendix contains a listing of the commercially available, concrete teaching devices that have been used to illustrate concepts throughout the book. Each listing contains a sketch or photograph of the device, a brief description or example of how it is used, and the mathematical topics usually taught with it. See the index for the pages of the textbook in which the use of each item is discussed in greater detail.

Photo A–1

Both the slide and loop abacuses are used to teach place value, especially as it relates to numeration and the addition and subtraction algorithms.

Example: Ask the child to display a two- or three-digit number on the loop abacus. 336 would look like the arrangement in Figure A–1.

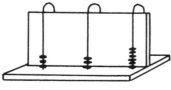

FIGURE A–1

Attribute Blocks

Photo A–2

Attribute blocks (or property blocks) are wooden or plastic blocks that vary in color, shape, thickness, size, or similar attributes that are easily observable by children. The blocks are used for classification and logic activities. Loops of plastic or heavy cord are used to enclose subsets of the blocks.

Example: Ask the child to sort the blocks so all the green ones are in one loop and the triangles in another (see Figure A–2). Discuss the placement of the green triangles with the child.

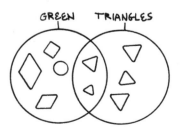

FIGURE A–2

Beam Balance

Photo A–3

The beam balance is a wooden or plastic beam, approximately 80 cm long, balanced on a central axis or fulcrum. Each side of the beam is divided into equal
units that are numbered one through ten from the center out to the end of the
side. At each unit, a peg is attached to the beam, allowing masses to be hung
under each numeral. A set of uniform masses of approximately 10 grams each
is provided.

Since the balance is an excellent model of an equation or inequality, it
can be used to demonstrate any arithmetic operation and clearly depicts the
commutative, associative, and distributive properties, as well as the reciprocal
nature of addition and subtraction or multiplication and division.

Example: Place masses on the 3 and 6 on one side. Ask the child to balance it with one mass on the other side. The child should place a mass on the
nine on the other side. Thus, $3 + 6 = 9$.

Next, place a mass on the 9 on one side and the 6 on the other side. Ask
the child what must be placed with the 6 to create a balanced beam. By placing
a mass on the 3 on the side with the 6, the child has shown that $9 = 6 + 3$.

Centimeter Cubes

Photo A–4

Centimeter Cubes are interlocking, plastic cubes in 10 colors. Each cube has a mass of 1 gram.

The cubes can be used to teach concepts of length, area, volume, or capacity, using metric measures. They also may be used for teaching fractions, number patterns, and geometric relationships.

Example: Fill a cubic decimeter with centimeter cubes. How many cubes does it take? A cubic decimeter is equivalent to a liter. How many cubic centimeters are there in a liter? What is the mass?

Photo A–5

The color cubes are brightly painted wooden cubes. The cubes may be used for counting, sorting, place value, basic operations, and activities in geometry and measurement.

Example: Set up a pattern using different colored blocks. Ask the child to copy or extend the pattern. Let children create their own two- or three-dimensional patterns.

Chip Trading

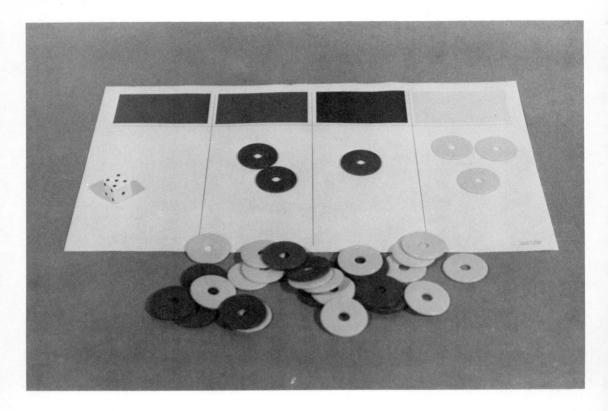

Photo A–6

The set of Chip Trading materials contains plastic chips in five colors, chip board, abacus board, numeral cards, operations cards, student books, and a teacher's guide.

Chip Trading may be used for developing place value concepts in any base. As well, the materials may be used to develop all the operations with whole numbers.

Example: Play a game of three-for-one trades. Three yellow chips are traded for one blue, three blues for one green, and three greens for one red. Children take turns rolling a die and taking the corresponding number of yellow chips. After each move, all possible trades are made. The first child to get one red wins.

Counting Chips

Photo A–7

Counting chips are round, plastic chips in a variety of colors. They may be used for teaching color recognition, counting, sorting, place value, and the basic arithmetic operations.

　　Example: Give the child three red chips and seven blue chips. Ask: "How many more red chips will you need to have as many reds as blues?"

Counting Sticks

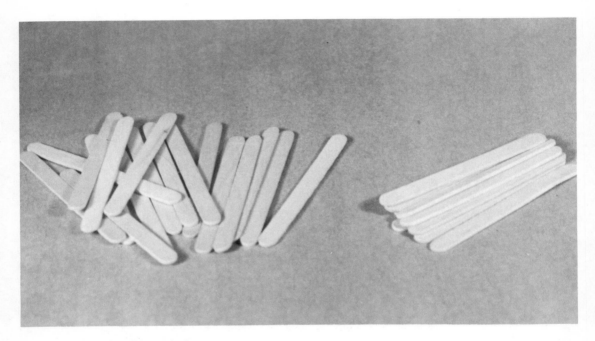

Photo A–8

Counting sticks are small, wooden sticks usually about 10 cm long and 2 cm wide. These sticks may be used to teach counting, place value, and the basic arithmetic operations.

Example: Give the child a pile of between 50 and 90 sticks. Ask him to tell you how many sticks there are. After the child has counted the sticks a few times, getting different answers each time, ask if there is an easier way to keep track of the sticks. Encourage the children to put sticks in equivalent piles or to put a rubber band around equal numbers of sticks. The child should discover that a pile or group of 10 is the most convenient group to count.

Cuisenaire Rods

Photo A-9

The Cuisenaire rods are wooden or plastic rods in 10 different colors that progress from 1 to 10 cm in length. All have a 1 cm² cross section. An important tool in any early childhood classroom, the rods may be used to demonstrate almost any arithmetic concept. Children can explore relationships between whole or fractional numbers using the rods. Rods may be used to demonstrate all basic operations. They may be used for such topics as equalities and inequalities, factors, fractions, proportion, sets, place value, and numeration.

Example: Choose a rod of any color. Ask the child to make as many trains that are the same length as the initial rod as possible. Some of the trains equivalent to a yellow rod are shown in Figure A-3.

y				
r		r		w
w	w	w	w	w
p				w
r		g		

FIGURE A-3

Fraction Bars

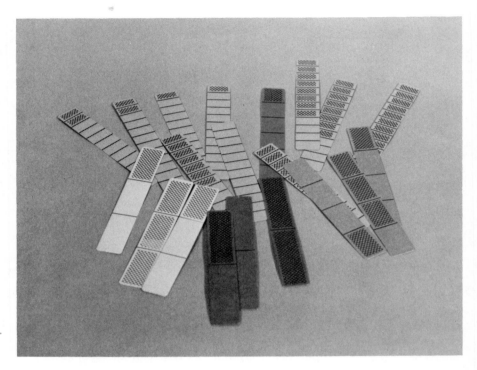

Photo A–10

The Fraction Bars are a set of materials in five colors divided into 2, 3, 4, 6, and 12 equal parts. Workbooks, game mats, spinners, dice, activity cards, playing cards, and teacher's guides accompany the set.

The bars offer a concrete introduction to fraction concepts. They may be used for a variety of fraction ideas including equivalent fractions and addition, subtraction, multiplication, and division of fractions.

Example: Given a bar such as in Figure A–4a, find another bar with the same amount of shading (see Figure A–4b). Are there any more bars with the same amount of shading? What can you say about 1/3 and 2/6?

FIGURE A–4a   **FIGURE A–4b**

Photo A–11

Geoblocks are pieces of unfinished hardwood cut into a wide variety of shapes and sizes. They may be used to help children develop intuitive ideas of surface area, volume, three-dimensional space, two-dimensional projections, scaling, and mapping.

Example: Put a block into a bag so the children may feel it but not see it. After feeling the block, ask the children to:

1. Pick out another block just like it,
2. Describe the block, so that another child may pick out a block just like it, or
3. Draw a picture of the block.

Geoboards

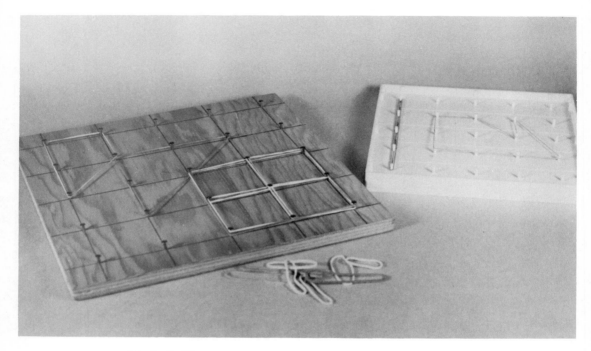

Photo A–12

Geoboards are plastic or wooden boards with nails or pegs in an array, typically 5 × 5. A clear plastic geoboard is available that may be used on an overhead projector.

Geoboards may be used for a variety of concepts, not all of which are classified as geometry. Some of these are number concepts and patterns, Cartesian coordinates, shape, area, length, sets, angles, symmetry, and fractions.

Example: Make a pattern on the geoboard using the overhead projector. Children should copy the pattern on their own geoboards.

Graduated Beakers and Cylinders

Photo A–13

Graduated beakers and cylinders are commonly plastic and vary in size from 10 ml to 1 liter.

The beakers and cylinders may be used for many types of measuring activities, including determining liquid amounts and the volume of solid objects, using the water displacement method. The relationship between 1 ml and 1 cm^3 may be found using this method.

Example: Put 20 ml of water in a graduated cylinder. Immerse 5 centimeter cubes in the water, and read the level of the water. How much did it increase? What is the relationship between 1 ml and 1 cm^3?

Meter Sticks and Metric Rulers

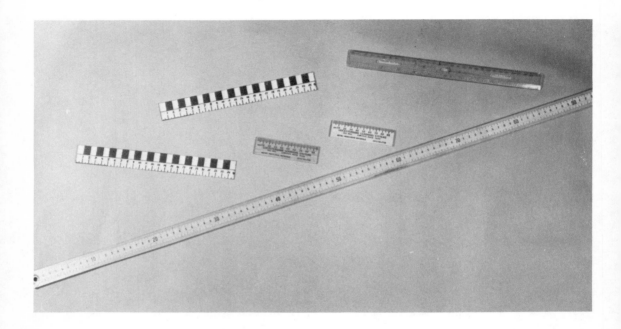

Photo A–14

Meter sticks are metric rulers one meter in length having markings for dm, cm, and perhaps mm. Thirty cm metric rulers are also available.

Meter sticks and shorter rulers are essential to linear metric measurement. Teachers may purchase commercial ones, or the students may make their own.

Example: Make a chart with the following headings: Less than 1 m, About 1 m, Greater than 1 m. Give each child a meter stick and let him or her fill in the chart using classroom objects.

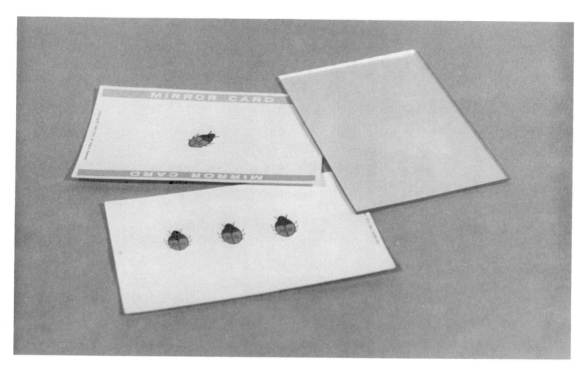

Photo A–15

The mirror cards set contains approximately fifteen packets of ten or twelve cards each. Each packet contains three types of cards: (1) an object card containing a simple pattern that can be made more complex by placing a mirror on edge adjacent to the pattern; (2) a set of cards depicting complex patterns that may be formed by placing the mirror appropriately on the object card; and (3) a set of cards depicting complex patterns that are impossible to form using the mirror on the object card.

Example: Give the child an object card and a set of pattern cards. Using the mirror on the object card, the child tries to match the image on the pattern card. The child sorts the pattern cards into "match" and "do not match" piles.

Multibase Blocks

Photo A–16

The Multibase Blocks are a set of wooden or plastic cubes showing base relationships. For example, if the base is ten, the long is ten times as long as the unit, the flat is ten times as big as the long, and the block is ten times as large as the flat. Similar relationships exist for sets of blocks depicting bases less than 10.

These blocks are excellent for building place value concepts. They may be used to demonstrate the concepts of addition, subtraction, multiplication, and division.

Example: Using the base ten longs and units, ask the children to do two-digit addition requiring regrouping. 18 + 36 may look like the arrangement in Figure A–5. The units are regrouped. Ten units are traded for a long, leaving 4 units. The result appears in Figure A–5.

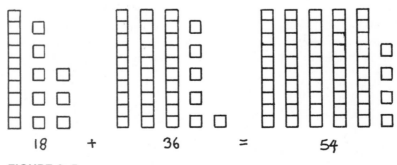

FIGURE A–5

Pan Balance

Photo A–17

The pan balance is a balance having pans on each end of the beam in which uniform masses or other objects may be placed. A set of standard masses ranging from one gram to five hundred grams are provided with the balance.

The pan balance may be used to teach the concept of mass measurement with either standard or arbitrary units.

Example: Put a small book in one of the pans and try to balance it with inch blocks.

Parquetry Blocks and Cards

Photo A–18

Parquetry Blocks are a set of wooden blocks of assorted colors and shapes but uniform thickness. The Parquetry Cards are pictures of patterns that may be made from the Parquetry Blocks.

The Parquetry Blocks are used to teach geometric concepts such as symmetry, congruence, similarity, area, and angles.

Example: Give the child a set of Parquetry Cards and the Parquetry Blocks. Tell the child to make the pattern with the blocks.

Photo A–19

The Pattern Blocks are wooden blocks in six colors and shapes having uniform thickness. The kit usually contains one or two metal mirrors.

Pattern Blocks can be used to investigate geometric forms and relationships such as symmetry, congruence, shapes, size, area, and angles. They may also be used for counting, sorting, and matching. The blocks are related in such a way that they may be used for work with fractions.

Example: Make a pattern with the blocks, and ask a child to copy or extend the pattern. Children may also make patterns for each other.

People Pieces

Photo A–20

The People Pieces are an attribute material consisting of 16 wooden or plastic blocks illustrated with "people" of 2 colors, 2 sexes, 2 weights, and 2 heights.

People Pieces may be used for a variety of activities to develop logical thinking skills, classification, and understanding of relationships.

Example: Ask the child to sort the blocks so that all the blocks that go together are piled together. Is there more than one way to do this?

Sequencing Beads

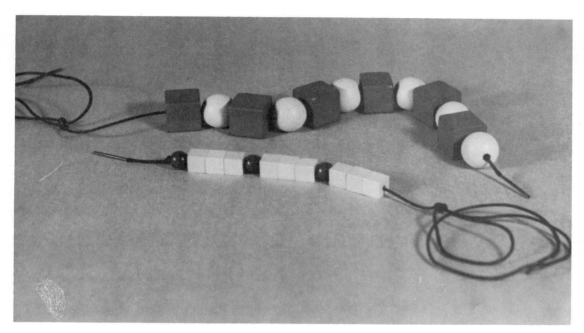

Photo A–21

Sequencing beads are wooden beads in a variety of shapes, sizes, and colors.

The beads may be used for sequencing activities that require the child to copy, remember, or create a pattern. They help develop visual memory and discrimination.

Example: Give the child a set of cards that show beads strung in different ways. Ask the child to string the beads to match the cards.

Stacking and Nesting Blocks

Photo A–22

Stacking and nesting blocks are commonly constructed of plastic or wood and are available in a variety of shapes, sizes, and colors. The parts of any set either stack or nest or both.

The blocks help children sequence materials by size. The activities are immediately self-checking, because the blocks can be stacked or nested in only one way. Use of the blocks may also help develop color discrimination, counting, and order relationships.

Example: Give the child a set of stacking blocks, and ask him to make a tower. The largest block should be placed on the bottom with the sizes decreasing as the tower is built.

Tangrams

Photo A–23

Tangrams are sets of seven wooden or plastic pieces containing five triangles of proportional sizes, a parallelogram, and a square.

This puzzle, originally invented by the Chinese thousands of years ago, may be used to challenge children to produce a variety of geometric shapes and other silhouette figures. Tangrams help develop spatial perception, fractional concepts, and problem solving.

Example: Take the large triangle. Can this form be made using any of the other shapes? Is there more than one way to do it?

Trundle Wheel

Photo A–24

The trundle wheel is a plastic or wooden wheel that gives an audible click on each revolution. Since the circumference of the wheel is one meter, the wheel travels one meter for each click.

The wheel is used to measure distances of several meters. It can be used to demonstrate the relationship between the diameter and circumference of a circle.

Example: Give the child a trundle wheel and ask him to measure such things as the perimeter of the room, the hallway, a track for the 50 meter dash, and other similar distances.

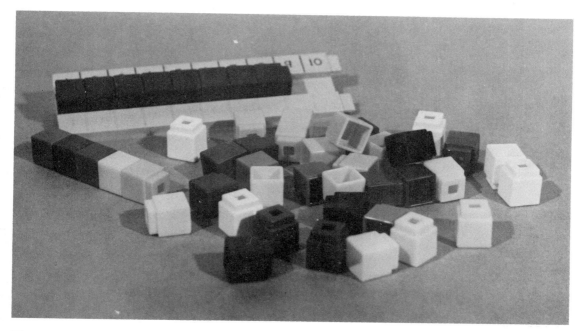

Photo A-25

The Unifix materials consist of interlocking cubes in 10 colors, 1 to 10 stairs, number indicators, 10 × 10 inset pattern boards, ten and units trays, a 100 track, grid work guide, and teacher's guide.

Unifix cubes may be used to demonstrate arithmetic ideas including number concepts, numeration, place value, and the four basic operations.

Example: Using the hundred track, put a yellow cube on every second number and a green cube on every third number. What patterns do you notice? Does any number have both a yellow and green cube? Does any odd number have a yellow and a green cube?

Suppliers of Manipulative Materials

Most of the materials listed in the appendix are available from several sources. Each of the following companies has some or all of the materials listed:

Creative Publications
P. O. Box 10328
Palo Alto, CA 94303

Cuisenaire Company of America
12 Church Street
New Rochelle, NY 10805

Dick Blick Co.
P. O. Box 1267
Galesburg, IL 61401

Didax
P. O. Box 2258
Peabody, MA 01960

ETA School Materials Division
159 East Kinzie Street
Chicago, IL 60610

O Haus Scale Corporation
29 Hanover Road
Florham Park, NJ 07932

Scott Resources, Inc.
1900 East Lincoln, Box 2121
Fort Collins, CO 80522

Selective Educational Equipment, Inc.
3 Bridge Street
Newton, MA 02159

Sigma Scientific, Inc.
P. O. Box 1302
Gainesville, FL 32601

Teaching Resources
100 Boylston Street
Boston, MA 02116

Index

A

Abacus, 87, 137, 140
  loop, 317
  slide, 317
Abstraction, 8, 42–43
Abstract level of cognition, 123–126
Active involvement, 1
Activity cards, 362
Addition
  algorithms, 137–140
  introduction, 115–117
Angle, intuitive concept, 176
Area, 185, 192–196, 203–206, 223–224
Arithmetic Teacher, 282
Arrays, 120
Art, children's drawings, 157
Ashlock, Robert B., 149
Association for Supervision and Curriculum Development, 28
Associative Property
  number operations, 126–130
  operations on sets, 112–114
Attribute blocks, 38, 42–43, 45, 53, 109, 113–115, 234–235, 238–240, 243, 246, 253–254, 257, 318
Audio-visual materials, 312

B

Balance
  beam, 319
  pan, 116, 120, 333
Baratta-Lorton, Mary, 69, 107
Base ten blocks, 137–138, 141

Basic number facts, 115–138
Beakers, graduated, 329
Bean sticks, 87, 90, 137, 139, 142
Beliefs about children, 15–17
Bell, Frederick H., 314
Biggs, Edith, 17, 28, 314
Bingo, 76
Bloom, Benjamin, 231, 273
Brownell, William A., 7, 28, 149
Bruner, Jerome S., 6, 28

C

Calculators, 146–147
Carroll Diagram, 245, 247–250
Cartesian cross product, 120–121
Centimeter cubes, 320
Chip trading, 87, 93, 137, 139, 142–143, 322
Classification, 31, 36–42, 237–240
  of objects, 36–42, 53–55
  recognizing, 31
  of sets, 56
Classroom management, 24–25
Color cubes, 33, 42, 116–117, 302, 321
Commercial materials, 297–362
Communicating:
  abstractions, 14–15
  comparative, noncomparative, 36
  describing objects, 31–36
  mathematical concepts, 11–15
Commutative property
  operations on number, 126–130
  operations on sets, 111–112
Comparing, problem solving, 236–237